T0297964

MAGNETICALLY INDUCED ELECTROMOTIVE FORCE

Frederick J. Young, Doctor of Philosophy
800 Minard Run Road, Bradford, Pennsylvania, US
15 October 2012

Michael Faraday, the Father of Emf

124229-YOUN

ISBN: Softcover 978-1-4797-3789-5

To order additional copies of this book, contact:
Xlibris Corporation
1-888-795-4274
www.Xlibris.com
Orders@Xlibris.com

Magnetically Induced Electromotive Force

Frederick J. Young, Doctor of Philosophy
800 Minard Run Road, Bradford, Pennsylvania, US

15 October 2012

Contents

Part I
PREFACE

As professor of electrical engineering more than a half century ago I taught advanced electromagnetic field theory at the Carnegie Institute of Technology along with my colleagues, Drs. Everard Williams, Benjamin Teare, R. Louis Bright, Edward Schatz and James B. Woodford to several classes of doctoral candidates. Most of the graduate students were rather confused about electromotive force and its calculation. The misconceptions they had learned from various textbooks as undergraduates involving flux cutting had to be discarded. Much of the matter in this book was first presented in the class lecture notes of Dr. R. Louis Bright inspired by the publications of and by association with Dr. Joseph Slepian, the assistant director of the Westinghouse Research Laboratories at East Pittsburgh, PA. Drs. Everard Williams, Benjamin Teare, Edward Schatz, James B. Woodford and Gaylord Penney, (inventor of the electrostatic precipitron and former director of the Westinghouse Research Laboratories at East Pittsburgh, PA.). In the 1949, 1950 and 1951 issues of the Transactions of the American Institute of Electrical Engineers several essays concerning the understanding and calculation of electromotive force were published[1]by Dr. Joseph Slepian. He contributed six essays entitled "Electrical Essays" and many answers to the problems he posed to the readers. There were also many discussions in various foreign journals concerning emf. These lecture notes were used and augmented by other professors in the electrical engineering department at Carnegie Institute of Technology in Pittsburgh, PA, also. The students had to work many of the problems contained herein. For several years after these giants quit teaching I presented this material to many classes of graduate students and added to the development of the subject. When I was consultant to the Westinghouse Research Laboratory in Churchill, PA, a Pittsburgh suburb, several of us would have heated discussions about emf. The participants included Drs. R. Louis Bright (the new assistant director), Clarence Zener (the new director of the Westinghouse Research Laboratory), inventor of the Zener diode and me. In these sessions one of us would bring up a new emf problem and everybody would try to find easy solutions. These meetings were most stimulating and productive. At present we have a much greater ability to make quick searches and find out what various people think about almost any subject. **I searched for electromotive force on the internet and found many sources. How-**

Figure 1: Michael Faraday in lab

ever, most of the sources have not yet learned the contents of this book and many are quite wrong[2,8], exhibiting no real understanding of emf. For that reason I publish this book and wish to acknowledge those who helped me to understand emf and collect the best of the subject. The preface to this work would not be complete without reference to the founding father of emf, Michael Faraday. It is most interesting that almost 150 years after his death, Faraday's Law about the generation of emf is still little understood!

1 Michael Faraday, the Father of Emf

Michael Faraday was born into humble conditions, brought up in the Sandemanian sect of the Christian Church and made his name in the scientific world, despite his lack of formal education, through his outstanding discoveries, observations and experiments. His scientific work laid the foundations of all subsequent electro-technology. From his experiments came devices which led directly to the modern electric motor, generator and transformer.

Michael Faraday was born on 22 September 1791. At the age of fourteen he was apprenticed to a London bookbinder. Reading many of the books in the shop, Faraday became fascinated by science, and wrote to Sir Humphrey Davy at the Royal Institution asking for a job.

On 1 March 1813, he was appointed laboratory assistant at the Royal Institution. There Faraday immersed himself in the study of chemistry, becoming a skilled analytical chemist. In 1823 he discovered that chlorine could be liquefied and in 1825 he discovered a new substance known today as benzene.

1.1 Electromagnetic Technology

However, his greatest work was with electricity. In 1821, soon after the Danish chemist, Oersted, discovered the phenomenon of electromagnetism, Faraday built two devices to produce what he called electromagnetic rotation: that is a continuous circular motion from the circular magnetic force around a wire.

Ten years later, in 1831, he began his great series of experiments in which he discovered electromagnetic induction. These experiments form the basis of modern electromagnetic technology.

On 29 August 1831, using his "induction ring", Faraday made one of his greatest discoveries - electromagnetic induction: the "induction" or generation of electricity in a wire by means of the electromagnetic effect of a current in another wire. The induction ring was the first electric transformer.

1.2 Magneto-electric Induction

In a second series of experiments in September he discovered magneto-electric induction: the production of a steady electric current. To do this, Faraday attached two wires through a sliding contact to a copper disc. By rotating the disc between the poles of a horseshoe magnet he obtained a continuous direct current. This was the first generator.

Although neither of Faraday's devices is of practical use today they enhanced immeasurably the theoretical understanding of electricity and magnetism. He described these experiments in two papers presented to the Royal Society on 24 November 1831 and 12 January 1832.

These were the first and second parts of his "Experimental researches into electricity" in which he gave his ,"law which governs the evolution of electricity by magneto-electric induction".

After reading this, a young Frenchman, Hippolyte Pixii, constructed an electric generator that utilized the rotary motion between magnet and coils rather than Faraday's to and fro motion in a straight line. All the generators in power stations today are direct descendants of the machine developed by Pixii from Faraday's first principles.

1.3 Later work

Faraday continued his electrical experiments. In 1832 he proved that the electricity induced from a magnet, voltaic electricity produced by a battery, and static electricity were all the same. He also did significant work in electrochemistry, stating the First and Second Laws of Electrolysis. This laid the basis for electrochemistry, another great modern industry.

Faraday's descriptive theory of lines of force moving between bodies with electrical and magnetic properties enabled James Clerk Maxwell to formulate an exact mathematical theory of the propagation of electromagnetic waves.

In 1865 Maxwell proved mathematically that electromagnetic phenomena are propagated as waves through space with the velocity of light, thereby laying the foundation of radio communication confirmed experimentally in 1888 by Hertz and developed for practical use by Nikola Tesla and Guglielmo at the turn of the century. For a long time Marconi was credited with the invention of radio. However, posthumously Tesla won a court judgement that he was the real inventor of radio.

1.3.1 Faraday the lecturer and educator

Faraday was also the greatest scientific lecturer of his day, who did much to publicize the great advances of nineteenth century science and technology through his articles, correspondence and the Friday evening discourses which he established at the Royal Institution. He considered it a vital part of his job to educate the public on cutting-edge science. The Royal Institution Christmas lectures for children, begun by Faraday, continue to this day.

1.3.2 Later life

In 1865 Faraday ended his connection with the Royal Institution after over 50 years of service. He died at his house at Hampton Court on 25 August 1867. His discoveries have had an incalculable effect on subsequent scientific and technical development. He was a true pioneer of scientific discovery.

1.4 Michael Faraday and the Institute of Engineering Technology (IET)

'Faraday' is a name of great significance in the IET. His statue stands outside Savoy Place, the London home of the IET, and his name is given to a major resource aimed at getting young people into science and engineering. We also hold an important collection of his personal papers in the IET Archives. But why is Michael Faraday so important to the IET? After all he died in 1867, four years before the Institution was formed.

Faraday's journey from unqualified apprentice to the most famous scientist of his day, his work on the fundamentals of physics and electricity and his untiring efforts to promote science in society have ensured his legacy to the history of science and technology and the importance of his name in promoting the IET today.

Frederick J. Young, Ph. D.

Part II
INTRODUCTION

There exists a great amount of confusion about the calculation of electromotive force commonly denoted as emf. A fallacious understanding of Faraday's law is given in many texts and in various internet sources. This short text is an attempt to present a clear set of rules for evaluating the electromotive force induced in a circuit as a consequence of either a time variation of a magnetic field or the movement of segments of the circuit through a magnetic field or both.

The introductory treatment of this problem as presented in most elementary texts is inadequate both because the quantities are not often carefully defined and because the restrictions upon the validity of the formulas given are not emphasized. It is hoped the treatment herein which these rules are rigorously derived from Maxwell's Equations and the Lorentz Transformation Equations will clarify the confusion so common in this subject.

Because of its relative simplicity and also because it illustrates several of the common pitfalls, the first section of this short text is restricted to the case in which there is no relative motion between the parts of the circuit about which the emf is being calculated. Such circuits are called 'rigid circuits'.

Part III
RIGID CIRCUITS

The specific problem to be considered in this section is the determination of the electromotive force induced in a rigid conducting circuit as a consequence of a time variation in a magnetic field. Before attempting the analysis of this problem it is necessary to carefully define what is meant by electromotive force. The definition used in this section is given by:

> **The instantaneous electromotive force about any closed rigid path is equal to the line integral of the instantaneous electric field strength integrated along that path.**

The equation for this is

$$e = \oint \mathbf{E} \cdot dl \tag{1}$$

where e = the instantaneous electromotive force (emf) about the prescribed path

dl = the incremental element of the path

$\mathbf{E}$ = the vector of electric field intensity at the position dl

The electric field intensity, $\mathbf{E}$, is understood to be referred to a reference frame which is stationary with respect to the rigid path about which the emf is to be evaluated. This field intensity is defined as the force per unit positive charge that an infinitessimal charge would experience if placed in a long needle-like cavity parallel to the force and located at the point at which the field was to be evaluated. $\mathbf{E}$ is the electric field intensity or strength that appears in Maxwell's equations.

2 Consistency with Kirchhoff's Voltage Law

It will be shown now that this definition of emf is consistent with the usual form of Kirchhoff's Voltage Law, given by:

The emf about any closed loop is equal to the sum of the IR drops about that loop. Consider the arbitrary configuration of finite conductors where the conductors are connected any imaginable way. They are all finitely thick and have bounding surfaces. If there are no eddy currents in the conductors and if the displacement currents, $\partial \mathbf{D}/\partial t$, due to the distributed capacitance are negligible, it is possible to divide any such circuit into a group of elements each of which is bounded on its sides by stream surfaces and on two ends by equipotential surfaces. The current through any element thus is constant throughout the element and the potential across the element is independent of the path through that element. Then one may break up the closed line integral into the sum of the line integrals over each such segment of the path. Then

$$e = \oint \mathbf{E} \cdot dl = \sum_i \int_i \mathbf{E} \cdot dl \tag{2}$$

where $\int_i \mathbf{E} \cdot dl$ denotes the line integral over the ith segment and $\sum_i$ indicates the summation over all the segments of the prescribed closed loop. The current in the ith segment is equal to $\int_i \mathbf{J} \cdot dS$ where $\mathbf{J}$ is the current density at the element of area $d\mathbf{S}$ and the integration is carried out over any cross section of the ith element. One may then multiply and divide each term of the preceding sum given by Eq.2 by this quantity to obtain

$$e = \sum_i \int_i \mathbf{E} \cdot dl = (\sum_i \int_i \mathbf{J} \cdot dS) \cdot [(\int_i \mathbf{E} \cdot dl) / \int_i \mathbf{J} \cdot dS] \tag{3}$$

If all the media containing the fields associated with this circuit are linear and $\mathbf{E}$ is increased by a factor k at any one point then it must increase by the factor k everywhere, and because for linear media, $\mathbf{J} \propto \mathbf{E}$, $\mathbf{J}$ must everywhere increase by the same factor. Therefore both the numerator and the denominator of the expression $[(\int_i \mathbf{E} \cdot dl) / \int_i \mathbf{J} \cdot dS]$ shall increase by the same factor and hence the fraction is independent of the field strength and is a function only of the geometry of the circuit. This fraction is the quantity defined as the resistance of the ith element, R_i. Hence, Eq.3 for the emf becomes

$$e = (\sum_i \int_i \mathbf{J} \cdot dS) \cdot [(\int_i \mathbf{E} \cdot dl) / \int_i \mathbf{J} \cdot dS] = \sum_i I_i R_i \tag{4}$$

which is Kirchhoff's Law.

3 Magnetically Induced Electromotive Force in Rigid Circuits

Now that the definition of emf has been shown to be consistent with Kirchhoff's Law we will return to the original problem of this section; namely the evaluation of the magnetically induced electromotive force in a rigid circuit. The field vector **E** measured in a reference system stationary with respect to this rigid circuit must satisfy the Maxwell Equation given by

$$\nabla \times \mathbf{E} = -\partial \mathbf{B}/\partial t \tag{5}$$

An expression for the emf about any prescribed path may be found by integrating both sides of Eq.5 over any two-sided surface **S** bounded by the path about which the emf is to be calculated. This becomes

$$\int_S (\nabla \times \mathbf{E})\cdot \mathbf{d}S = -\int_S (\partial \mathbf{B}/\partial t)\cdot dS \tag{6}$$

By Stokes Theorem the left side Eq.6 becomes $\int_S (\nabla \times \mathbf{E})\cdot \mathbf{d}S = \oint \mathbf{E}\cdot dl$ which is the electromotive force or emf. Because the path of integration has been specified as rigid, the surface **S** is independent of time and the partial differential operator on the right side may be taken outside the integral to yield

$$e = -d/dt \int_S \mathbf{B}\cdot d\mathbf{S} \tag{7}$$

The surface integral, $\int_S \mathbf{B}\cdot d\mathbf{S}$ is by definition equal to the magnetic induction flux Φ linking the prescribed path (including that due to any current in the circuit itself) so that the equation becomes the familiar form of Faraday's Law

$$e = -d\Phi/dt \tag{8}$$

Note that the electromotive force has been defined only for a closed path (a mathematical line) and not for a closed circuit or a closed conductor, for is a finite conductor contains eddy current, the emfs evaluated about different paths will not in general be equal. The emf about a conducting circuit containing finite conductors is independent of the path through the conductors only if there are no eddy currents, i.e. if the field inside the conductors is irrotational. This definition thus avoids any necessity for the concept of partial flux linkages.

It should be emphasized that the evaluation of the flux Φ must be carried out by integrating over any two-sided surface and not over a one-sided surface as depicted in Fig.2[3], called a . If a magnetic flux is concentrated in a region such as an iron core and the circuit consists of N turns about this core; then,

Figure 2: Möbius strip unsuitable for evaluation of magnetic flux

on the assumption that the leakage flux is negligible, it is evident that the total flux Φ linking the circuit is equal to N times the flux that would link a circuit having only one turn about the core. Calling this latter flux Φ_{turn} one writes the familiar form

$$e = -Nd\Phi_{turn}/dt \tag{9}$$

Attention should be called to the very important fact that in any region in which there is a time varying magnetic field the electric field is not irrotational because $\nabla \times \mathbf{E} \neq \mathbf{0}$.Hence it is in general impossible to define a unique potential difference $V = -\int_A^B \mathbf{E} \cdot dl$ between two points in such a region. For example, note that the reading of a voltmeter used to measure the potential difference between points A and B in Fig.3 depends on the position of the voltmeter and its connecting leads.

However, it is possible to uniquely define a terminal voltage for such a circuit if the terminals are brought out into a region in which there is no varying magnetic field and if it is understood that the path by which this circuit is completed lies entirely within this irrotational region. For example when we speak of the terminal voltage of a transformer secondary, we tacitly assume the

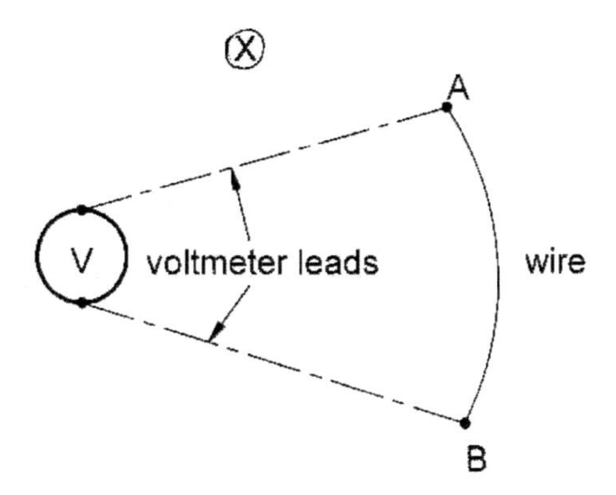

Figure 3: Potential depends on lead location

circuit connected to these terminals will not pass back through the transformer. Because $\partial\mathbf{B}/\partial\mathrm{t} = \mathbf{0}$ within this external region, the value of the surface integral $\int_S \partial\mathbf{B}/\partial\mathrm{t} \cdot \mathbf{dS}$ will be independent of the manner in which the circuit is completed and hence the emf may be calculated by assuming any convenient line between these terminals which will complete the boundary through this region. The potential V_{AB} between terminals A and B is defined as the line integral $-\int_A^B \mathbf{E} \cdot dl$ taken along any line (lying wholly within this irrotational region) which joins points A and B. The electromotive force, e, has been shown to be equal to $\oint \mathbf{E} \cdot dl$ taken about the closed circuit. Hence the terminal voltage V_{AB} is equal to $e - \int_A^B \mathbf{E} \cdot dl$ where the line integral in this cases is taken from A to B along the path through the transformer or generator and the line integral $\oint \mathbf{E} \cdot dl$ is taken as positive in going from the negative terminal A through the electrical machine to the positive terminal B and then back to A along a path in the irrotational region. Since the circuit in the generator has been assumed to be metallic, one may multiply and divide $\int_A^B \mathbf{E} \cdot dl$ by $\int_S \mathbf{J} \cdot d\mathbf{S}$ taken over the cross section of the conductor to yield

$$(\int_S \mathbf{J}{\cdot}d\mathbf{S})(\int_A^B \mathbf{E}{\cdot}dl)/\int_S \mathbf{J}{\cdot}d\mathbf{S} \qquad (10)$$

or in other words, the terminal voltage of a generator or transformer is equal to the induced voltage minus the IR drop in the generator. Then terminal voltage is given by $V_{AB} = e - \int_A^B \mathbf{E} \cdot dl$.

The section of the circuit external to this generator may contain any combination of resistors, batteries, capacitors, etc., and also the terminals of other such generators or transformers, provided that these other emfs and IR drops are properly taken into account when writing Kirchhoff's Voltage Law about the circuit. Note that because this law is written in the form $\sum e = \sum IR$, all circuit elements other than resistors must be spoken of as having an emf. For example, a battery, condenser or thermocouple are all said to contribute to the total emf where this contribution is positive if it is in the direction to aid the flow of current in the direction that was assumed positive in taking the line integral $\oint \mathbf{E}{\cdot}dl$, and negative if it opposes this flow of current.

If the circuit internal to the generator (e.g. in the rotational region) consists of anything other than a continuous, thin conductor, local circulating currents may be set up within the generator itself thus making the calculation of the terminal voltage very difficult. In such cases it is usually necessary to obtain the complete field solution in this region. Even in the case of a continuous, thin conductor it may be necessary in some cases to account for the displacement currents due to distributed capacitance.

Part IV
NONRIGID CIRCUITS

4 Sources of Electromotive Force or Emf

4.1 Electron Beam

Many elementary texts state that the emf generated in a conductor moving through a magnetic field is given by $e = Blv$. This is somewhat ambiguous because both **B** and **v** are relative vector quantities and the reference system in which they are measured must be specified. To illustrate this, consider a very thin beam of electrons moving with a velocity **u** relative to the electron gun as illustrated in Fig.4.

An observer at the point P who is stationary with respect to the electron gun shall see the electrons moving past him with a velocity u_z. The moving electrons constitute a current in the -z direction of λu_z where λ is the charge density per unit length in the electron beam. The application of Ampere's Law,

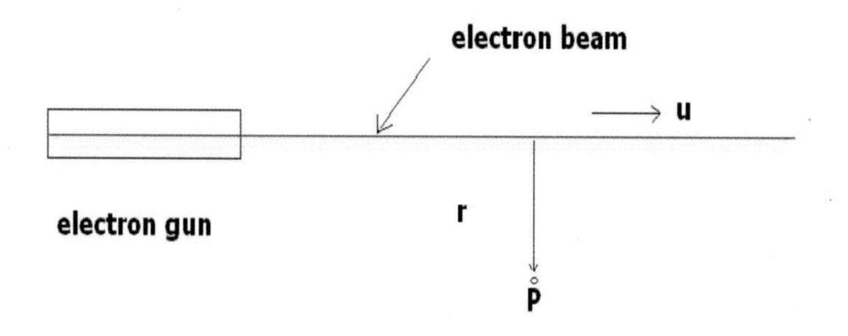

Figure 4: Moving electrons forming a beam

$\oint \mathbf{H}\cdot \mathbf{d}\ell = \int_S \mathbf{J}\cdot\mathbf{dS}$ is applied to a circular path centered on the electron beam. Then $\oint \mathbf{H}\cdot\mathbf{d}\ell = \mathbf{H}_\theta 2\pi r$. Because $\int_S \mathbf{J}\cdot\mathbf{dS} = \lambda u_z$ it follows that $\mathbf{H}_\theta = \lambda u_z/2\pi r$. Thus said observer shall be able to detect a magnetic field $B_\theta = \mu_0 \lambda u_z/2\pi r$ where $\mu_0$is the permeability of free space, and r is small enough that end effects can be neglected. However, let there exist another observer who is moving with a velocity of v_z (u_z and v_z are much smaller than the velocity of light). He will see the electrons moving past with a velocity of only $(u_z - v_z)$ and hence would calculate the magnetic field as $B_\theta = \mu_0\lambda(u_z - v_z)/2\pi r$. This example demonstrates that there is no unique magnetic field at any point in space, but rather the value of the magnetic field at any point in space may be different for two observers moving relative to one another. Here the observer stationary with respect to the electron gun is able to detect both and electric and a magnetic field, whilst a second observer moving parallel to the first with the same velocity as the electron stream would be able to detect only an electric field. This shows that if one is to apply any formula containing B_θ such as e = Blv to a system whose parts are in relative motion, one must in general specify the reference system B_θ is to be measured.

4.2 Current Carrying Wire

If, in the above case, the electron stream is replaced by a current carrying conductor, the situation is quite different. An observer stationary with respect to the wire sees a negative line charge density, $-\lambda$ moving to the right in Fig.5 with a velocity u_z and an equal positive charge density which is stationary. Because the total volume charge density in the conductor is always zero, the fixed charge has a net charge density of $+\lambda$ which appears stationary to this

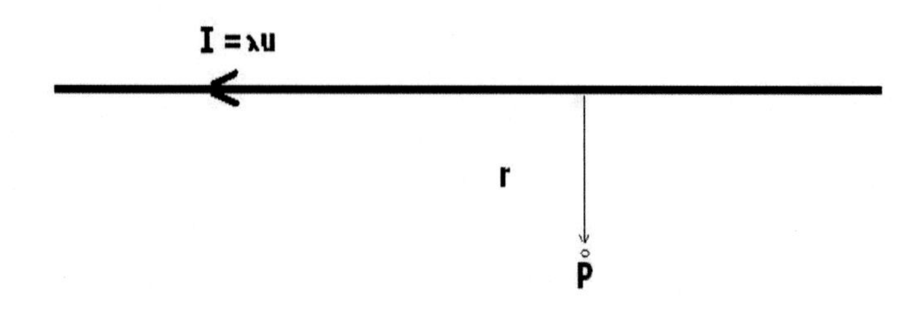

Figure 5: Current carrying wire

observer. The electric fields due to these two superimposed line charges cancel and only the moving charge produces a magnetic field. This magnetic field derivable from Ampere's Law is given also by $B_\theta = \mu_0 \lambda u_z / 2\pi r$.

An observer moving parallel to the wire with a velocity v_z will see a charge density of $-\lambda$ moving in the x direction with a velocity $(u_z - v_z)$ and also a charge of density $+\lambda$ moving in the -x direction with velocity of $-v_z$. The moving negative charge gives rise to a magnetic field in negative θ direction with a magnitude of $\mu_0 \lambda (u_z - v_z)/2\pi r$, whilst the moving positive charge produces a field of $\mu_0 \lambda v_z / 2\pi r$ also in the negative θ direction because both polarity of the charge and the direction of movement are reversed. The total field is then the sum of these two or $B_\theta = \mu_0 \lambda u_z / 2\pi r$.

To generalize, it may be shown from the special theory of relativity that if a certain observer sees the fields **B** and **E** at a given point in space, then the magnetic field at this same point as measured by a observer moving relative to the first with a velocity **v** will be given by

$$\mathbf{B}' = \mathbf{B} - (\mathbf{v} \times \mathbf{E})/c^2 \tag{11}$$

where c is the speed of light in vacuum which is 2.997925x10^8 meters per second. Thus, if as in the preceding example there is some observer in whose reference frame the electric field vanishes, than all observers will observe the same magnetic field. It follows that the magnetic field produced by a permanent magnet will also be independent of the velocity of the observer. It was shown experimentally by S. J. Barnett[17] rotating permanent magnets or solenoids about the axis of the magnetic field does not change the magnetic field. In other words, magnetic field lines do not rotate just because their source does.

4.3 Cutting Lines of Flux

Because the calculation of emf is greatly simplified in the magnetic field or induction is independent of the observer and because in the vast majority of practical applications the magnetic fields are produced either by current carrying conductors or by permanent magnets and thus possess this property, the remainder of this part of text will be restricted to such cases.

In this analysis a magnetic field is not regarded as an entity to which a velocity may be assigned. For example, if a permanent magnet is moving away from an observer, we shall not say that the observer sees the magnetic field moving along with the magnet but rather we shall say that this observer sees a magnetic field that is varying with time at all points in space. By saying that a conductor is moving through a magnetic field, we shall mean that the conductor is moving relative to some observer who also sees a magnetic field in the region of the conductor. However, we may not speak of the velocity of the conductor relative to the field nor do we speak of the conductor *"cutting lines of flux", a fruitful source of errors* . Similarly when a alternating current flows in a coil one should not think of lines of flux expanding and collapsing and thus *"cutting the conductors"*, but rather one should visualize a magnetic field which at all points is varying with time. The *"line of force"* as conceived by Faraday helps greatly in picturing many physical phenomena but in some cases such a concept is a definite disadvantage. *In the calculation of emf the engineer is far better off if he does not attempt to attribute any physical reality to such "lines".*

In calculating the emf about a circuit whose parts have relative motion one simply cannot use the form $\oint \mathbf{E}{\cdot}d\ell$ without further investigation, for the electric field $\mathbf{E}$ is also a relative quantity and the question again arises as to what frame of reference is to be used. To illustrate this relative property of $\mathbf{E}$ let us assume that a given observer sees both an electric field $\mathbf{E}$ and a magnetic field $\mathbf{B}$ at some given point in space. If he were to measure the force on a charged particle which was stationary with respect to him, he would obtain a force $\mathbf{F} = q\mathbf{E}$ from the definition of $\mathbf{E}$. However, if the charged particle were moving with a velocity $\mathbf{v}$ with respect to him and he again measured the force by observing the trajectory or some other means, he would find that $\mathbf{F} = q(\mathbf{E} + \mathbf{v} \times \mathbf{B})$. In other words the electric field as measured by an observer moving along with the charge would be $\mathbf{E}' = \mathbf{E} + \mathbf{v} \times \mathbf{B}$. This expression enables one to calculate the electric field in any frame of reference if the values of $\mathbf{E}$ and $\mathbf{B}$ are known in some other frame of reference. Again, this is a demonstration of the fact that the electric field and the magnetic induction are not independent of each other. This shows that the decomposition of an electromagnetic field into electric and magnetic components is not unique but depends upon the velocity of the observer.

5 Summary

Let $\mathbf{E}$ and $\mathbf{B}$ be the values of the electric intensity and magnetic induction fields as measured by some given observer and let $\mathbf{E}'$ and $\mathbf{B}'$ be the values measured by an observer moving with a velocity **v** relative to the first observer. Here $|\mathbf{v}| \langle\langle c$ the velocity of light. The relationships between these two set of values are given by

$$\mathbf{E}' = \mathbf{E} + \mathbf{v} \times \mathbf{B} \quad (12)$$

$$\mathbf{B}' = \mathbf{B} - (\mathbf{v} \times \mathbf{E})/c^2 \quad (13)$$

$$\mathbf{J}' = \mathbf{J} - \rho\mathbf{v} \quad (14)$$

From Eq.13 and Eq.14 it is clear that the magnetic induction and current density are almost unchanged by the relative velocities of the observers. The current densities are identical because no free charge exists in the wire. These equations are special cases of the equations derived later valid for $|\mathbf{v}| \langle\langle c$. In particular see Eqns. 71, 72 and 73.

Part V

Relative Motion between Circuit Parts

Keeping the cautions of the preceding paragraphs in mind, we shall develop a definition for the electromotive force about any closed metallic circuit whose various parts may be moving relative to each other and which may or may not contain sliding contacts. Once again the desired definition of emf must be consistent with Kirchhoff's Law. When there is relative motion between parts of the circuit the resistance of a segment must be defined in such a way that it does not change if the segment is moving. That is, if one wishes to retain the formula for the dissipation in an element as I^2R and because both the dissipation and the current are independent of the observer, it follows that the resistance to be used in this formula should be the same as the resistance calculated by an observer stationary with respect to that element. In that case the resistance of any element is defined as

$$R_i = (\int_i \mathbf{E}' \cdot dl)/(\int_i \mathbf{J}' \cdot \mathbf{dS}) \quad (15)$$

where the primes denote that the fields are measured by an observer at rest with respect to the conductor at the point dl. (if the conductor is flexible it can be considered as comprising an infinite number of increments each of which has a resistance that is evaluated by an observer on that particular increment.) Because the current flowing through a conductor is independent of the observer, the IR drops about the circuit can be written as

$$\sum_i I_i R_i = \left\{ \sum_i \left(\int_i \mathbf{J}' \cdot \mathbf{dS} \right) / \left[\left(\int_i \mathbf{E}' \cdot dl \right) / \left(\int_i \mathbf{J}' \cdot \mathbf{dS} \right) \right] \right\} = \sum_i \int_i \mathbf{E}' \cdot dl = \oint \mathbf{E}' \cdot dl \tag{16}$$

It thus appears that a reasonable definition of the electromotive force about any prescribed path in a non-rigid, conducting circuit is given by

$$e = \oint \mathbf{E}' \cdot dl \tag{17}$$

where $\mathbf{E}'$ is defined as the electric field intensity evaluated at the element dl by an observer stationary with respect to the conductor at that point. To avoid having to use several reference frames in evaluating Eq.17 , Eq.12 previously discussed can be used to transform each element of Eq.17 to single given reference system. The fields measured by an observer in this fixed reference frame will be denoted by the unprimed quantities. By solving Eq.12 for $\mathbf{E}$ and substituting it into Eq.5 there results

$$\nabla \times \mathbf{E}' = -\partial \mathbf{B}/\partial t + \nabla \times (\mathbf{v} \times \mathbf{B}) \tag{18}$$

Integrating both sides of Eq.18 over the surface enclosed by the path about which the emf is to be calculated and applying Stokes Theorem there results

$$e = -\int_S \partial \mathbf{B}/\partial \mathrm{t} \cdot \mathbf{dS} + \oint \mathbf{v} \times \mathbf{B} \cdot dl \tag{19}$$

where $\mathbf{B}$ is the total magnetic induction field due to all causes and has been assumed to be independent of the observer.

$\mathbf{v}$ is the velocity of the segment of the conductor at the point corresponding to dl *relative* to some given observer. (Note carefully that $\mathbf{v}$ is the velocity of the conductor and not the velocity of the path element dl in case the two differ as will be shown in later examples.)

$\partial \mathbf{B}/\partial \mathrm{t}$ is the vector time rate of change of $\mathbf{B}$ relative to the same observer.

Warning!! This formula has been derived from the results of the Special Theory of Relativity and hence is strictly justified only for the case of uniform motion. However, it is found to hold with negligible error for all accelerations that are physically attainable in the terrestrial laboratory.

Once again the expression for the electromotive force (emf) as exhibited in Eq.19 contains a surface integral and hence can be evaluated only for a closed circuit. It is nevertheless possible to define a unique potential difference between the terminals of a generator, just as it was in the case of rigid circuits. These terminals and the external circuit must lie within a region that to some observer is irrotational and that observer must see no part of the external circuit moving through a magnetic field. In other words, if it is possible to connect these terminals with a rigid line so that there is no emf in any closed circuit formed by this line and any permitted external circuit, then it is possible to uniquely define a terminal voltage which is equal to the emf (computed about the internal circuit and this rigid line) minus the IR drop in the generator: e.g.

$$V_{AB} = -\int_{A}^{B} \mathbf{E} \cdot dl \qquad (20)$$

As before the sign convention is such that the line integral of Eq.17 is taken as positive in the direction going from the negative terminal, A, through the generator to the positive terminal, B, and $\int_{A}^{B} \mathbf{E}' \cdot dl$ 'is taken from A to B through the generator.

This section is restricted to cases where any segment of the circuit moving relative to any observer through a magnetic field is a conductor. The case of a dielectric moving through a magnetic field will be presented later.

6 Examples using Eq.19 to determine the electromotive force

Before going further with this analysis it seems advisable to illustrate the applications of these formulae for the electromotive force by applying it in several examples. Various other expressions will derived and pitfalls pointed out in the discussions these examples. In all of the following examples it is assume that the currents flow in the circuit about which we are calculating the emfs are either zero or so small that the field due to them is negligible. In cases where that is not true a complete field solution must be used.

6.1 A short moving along subway rails next to the power conductor

Consider the system shown in Fig.6 in which all conductors are reduced to thin straight wires. A voltmeter is connected to two straight wires which run parallel to the conductor carrying a constant current, I. The two straight wires have a shorting bar across them moving away from the voltmeter with velocity **v**. The voltmeter is measuring the emf caused by the motion of the shorting bar through the azimuthal field of the current carrying wire.

The emf in this circuit (taken positive in the direction shown) could, for example, be calculated from the point of view of an observer stationary with respect to the voltmeter. To this observer the magnetic field at all points in space is invariant with time so that $\partial \mathbf{B}/\partial t = \mathbf{0}$ and hence the first integral of Eq.19 is zero. The velocity zero for all parts of the circuit except the shorting conductor having velocity v as depicted in Fig.6. By Ampere's Law the magnetic induction is $B = \mu_0 I/2\pi r$ and is directed into the paper; $\mathbf{v} \times \mathbf{B}$ therefore is directed along negative dl and $\mathbf{v} \times \mathbf{B} \cdot dl = -vBdl$. Then the emf is given by

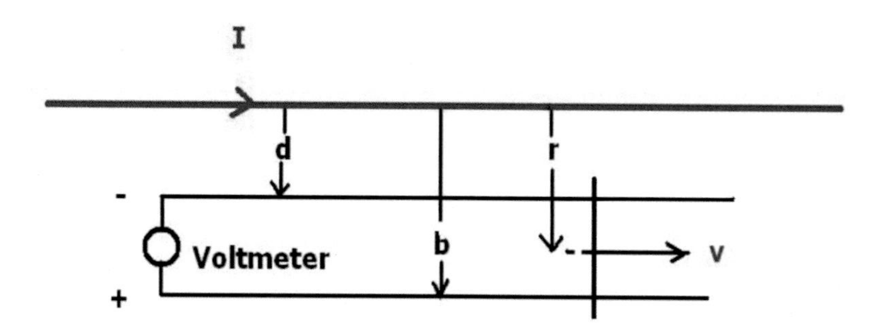

Figure 6: Example 1 Rail emf

$$e = -\int_{d}^{b} vBdr = -(\mu_0 Iv/2\pi)\int_{d}^{b} dr/r = -(\mu_0 Iv/2\pi)\cdot \ln(b/d) \qquad (21)$$

This emf might just as well have been calculated by assuming an observer fixed relative to the shorting bar. To this observer $\partial\mathbf{B}/\partial\mathbf{tT}$ would also have been zero and the first integral would disappear also. The velocity of the shorting bar relative to this observer is zero so that the second integral would vanish in the shorting bar. The rest of the circuit would have a velocity of v directed to the left. In the upper and lower conductors **v** and dl are parallel and therefore $\mathbf{v}\times\mathbf{B}\cdot dl = 0$. For this observer over the left end of the circuit where the voltmeter is, $\mathbf{v}\times\mathbf{B}$ is still in the opposite direction of dl. Thus Eq.19 yields the same emf as was obtained by the other observer. This example clearly illustrate that one must be very careful in speaking of the emf between two points. In this case the first observer claims the line integral $\oint \mathbf{v}\times\mathbf{B}\cdot dl$ has a value over the shorting bar whilst the other observer finds it to be zero. As a matter of fact, even in this situation in which there is no time varying field, the most Maxwell's Equations tell us is that $\oint \mathbf{E}'\cdot dl = \oint \mathbf{v}\times\mathbf{B}\cdot dl$. There is no justification for concluding from this that the integrands are equal or that the integrals are equal over a line segment. In other words, two different observers each using the formula for emf as given here may disagree as to the seat or source of the emf, but they will always agree on the value of the total emf about any closed circuit and it is only the total value that is of any use in computations.

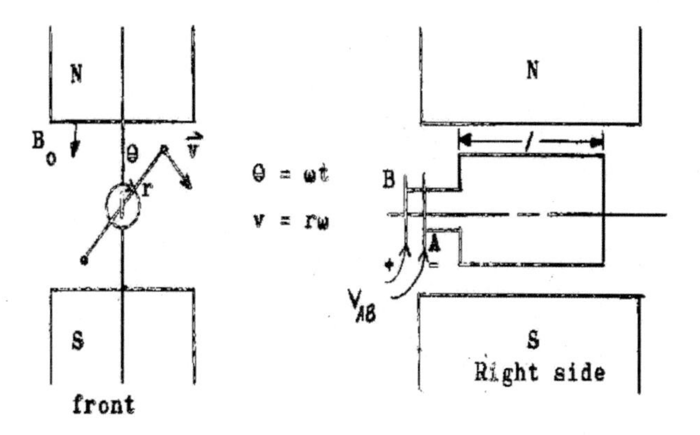

Figure 7: Simplification of a standard alternator

6.1.1 The terminal voltage

If the voltmeter has an internal resistance R_v and the rest of the circuit has a resistance of R, Eq.20 can be used to find the terminal voltage, V_{AB}. Then

$$V_{AB} = e - iR \tag{22}$$

and because $V_{AB} = iR_v$ also, where i is the current in the circuit of the voltmeter and shorting bar. Eq.22 becomes

$$iR_v = e - iR \tag{23}$$

whence

$$i = e/(R + R_v) \tag{24}$$

Substituting Eq.21, Eq.23 and Eq.24 into Eq.22 yields

$$V_{AB} = -\{\mu_0 Iv/[2\pi(1 + R/R_v)]\} \cdot \ln(b/d) \tag{25}$$

and the current becomes

$$i = -\{\mu_0 Iv/[2\pi(R + R_v)]\} \cdot \ln(b/d) \tag{26}$$

Here it is interesting to note the ratio of the open-circuit voltage to the short-circuit current is equal to R, which is the internal resistance of the generator.

6.2 Simplified standard alternator

For this example a simplification of a standard alternator consisting of a rectangular loop rotating in a constant, uniform magnetic induction field shall be considered.

If the observer is fixed relative to the poles, he sees no variation of $\mathbf{B}$ with time requiring $\int_S \partial\mathbf{B}/\partial t \cdot d\mathbf{S} = \mathbf{0}$, thus reducing the expression for the emf to $e =$

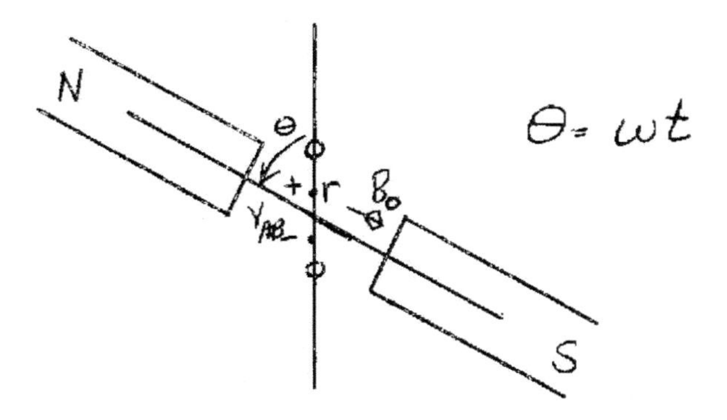

Figure 8: Fixed coil and rotating alternator poles

$\oint \mathbf{v} \times \mathbf{B} \cdot dl$. On the ends of the coil $\mathbf{v} \times \mathbf{B}$ is perpendicular to dl and therefore $\mathbf{v} \times \mathbf{B} \cdot dl = 0$ along these segments. Along the coil sides, $\mathbf{v} \times \mathbf{B} = B_0 v \cdot \cos(\theta)$ and is parallel to dl and on the upper side is directed opposite to dl. Thus

$$e = \int \mathbf{v} \times \mathbf{B} \cdot dl = -2B_0 lv \cdot \cos(\theta) \tag{27}$$

Because $\theta = \omega t$ and $v = r\omega$ when there is no current in the coil, the terminal voltage is given by

$$V_{AB} = -2B_0 lv \cdot \cos(\theta) = -2B_0 lr\omega \tag{28}$$

However, if the observer is taken as stationary with respect to the coil, the evaluation is entirely different. As far as this observer is concerned, no segment of this circuit that lies in the magnetic field is moving. Hence he believes $\oint \mathbf{v} \times \mathbf{B} \cdot dl = 0$. This observer does see a varying magnetic field. Its magnitude remains constant whilst it changes direction as shown in Fig.8 showing the coil fixed and the alternator poles rotating.

Neglecting fringing at any point on the coil $\mathbf{B} = \mathbf{i}B_0 \sin(\omega t) - \mathbf{j}B_0 \cos(\omega t)$ where $\mathrm{d}\mathbf{S} = \mathbf{i}\mathrm{dS}$ in accordance with the standard rules of vector analysis. Therefore, remembering that the area of the coil through which the magnetic induction passes is $2rl$

$$e = -\int_S \partial \mathbf{B}/\partial \mathrm{t} \cdot \mathbf{dS} = -\omega \int [\mathbf{i}B_0 \cos(\omega t) + \mathbf{j}B_0 \sin(\omega t)] \cdot \mathbf{i}dS$$

$$= -\omega B_0 \cos(\omega t) \int dS = -2B_0 lr\omega \cos(\omega t) \tag{29}$$

which is the same result the first observer found.

This example illustrates that the frequently used terms *motional emf* and *transformer emf* corresponding to $e = \oint \mathbf{v} \times \mathbf{B} \cdot dl$ and $e = -\int_S \partial \mathbf{B}/\partial t \cdot d\mathbf{S}$ are *meaningless unless the observer is specified.* In this example the one observer saw only a motional emf whilst the second saw only a transformer emf. However, it should not be concluded that an observer can always be chosen such that the emf is either purely transformer or purely motional.

6.3 Fallacious commutatorless direct current generator

If one were to use the method put forth in many texts for calculating the emf of the generator of Example 2 one might think a commutatorless generator is possible. It involves the application time and is reversed synchronously with the rotation of the coil so that the sides of the coil are always *cutting* flux in the same direction. Then one should be able to obtain a D. C. component from the generator. This proposal is now examined by Eq.19. The only difference between this and Example 2 is $B_0 = B_m \cos(\omega t)$ instead of a constant. As far as the secondary observer is concerned, there is still no motion of the coil and $\oint \mathbf{v} \times \mathbf{B} \cdot dl = 0$. He sees

$$\mathbf{B} = \mathbf{i}B_0 \sin(\omega t) - \mathbf{j}B_0 \cos(\omega t) = \mathbf{i}B_m \sin(\omega t)\cos(\omega t) - \mathbf{j}B_m \cos^2(\omega t) \qquad (30)$$

Using double angle trigonometry identities Eq.30 becomes $\mathbf{B} = \mathbf{B}_m\{\mathbf{i}\sin(2\omega t) - \mathbf{j}[1 + \cos(2\omega t)]\}/2$. Because the emf is given by $-\int_S \partial \mathbf{B}/\partial t \cdot d\mathbf{S}$ carrying out the indicated surface integral results in

$$e = -B_m\omega \int [\mathbf{i}\cos(2\omega t) + \mathbf{j}\sin(2\omega t)] \cdot \mathbf{i}dS = -2B_m lr\omega \cos(2\omega t) \qquad (31)$$

Thus this observer obtains a double frequency but no direct current component that a flux cutting calculation might lead one to expect. The first observer, stationary with respect to the poles, now sees both a moving coil and a varying magnetic field so that he must use both terms in Eq.19. The second term is evaluated as in Example 2 except now $B_0 = B_m \cos(2\omega t)$. There results

$$\oint \mathbf{v} \times \mathbf{B} \cdot dl = -2B_m lr\omega \cos(2\omega t) = -B_m lr\omega[1 + \cos(2\omega t)] \qquad (32)$$

If one were to stop with only this term, it would appear as if there were a D. C. component. However, the first term of Eq.19 must be examined. The partial derivative of $\mathbf{B}$ with respect to time as evaluated by this observer is $-B_m\omega \sin(\omega t)$ and points in the same positive direction as $\mathbf{B}$. Thus

$$-\int_S \partial \mathbf{B}/\partial t \cdot d\mathbf{S} = -\int (\partial \mathbf{B}/\partial t) \sin(\theta) \cdot dS = B_m\omega \int \sin^2(\omega t) dS \qquad (33)$$

$$-\int_S \partial \mathbf{B}/\partial t \cdot d\mathbf{S} = 2B_m lr\omega \sin^2(\omega t) = B_m lr\omega[1 - \cos(2\omega t)] \qquad (34)$$

The total emf is the sum of Eq.32 and Eq.34 which is given by

$$e = -2B_m lr\omega \cos(2\omega t)$$

In conclusion both observers see *no D. C. component of emf.* Reflection upon the example shall indicate it to be impossible to choose an observer for whom the transformer component of the emf vanishes and that one must be very careful with the directions of the various vectors.

6.4 Faraday disc generator

It is evident from the theory and as been shown by the preceding examples, the value obtained for the terminal voltage of a generator must be independent of the choice of observer. Because this is true and in many cases it is much simpler for one particular observer to evaluate the required integrals than it is for any other, we will not attempt to consider all the reasonable stations for the observers. Instead only the position of the observer who can make the simplest analysis is considered.

This example is the Faraday disc generator which comprises a circular, metal disc rotating in, and perpendicular to a uniform, constant magnetic induction field and having a brush on its axis and another on its periphery[4] .Shown in Fig.9 is a drawing of a Faraday disc, the first electromagnetic generator, invented by British scientist Michael Faraday in 1831. The copper disc rotated between the poles of a horseshoe shaped magnet. The motion induced a radial flow of current in the disk. The current flows into the spring contact sliding along the edge of the disk, through an external circuit, and back into the disk through the axle. The labeled parts are given in the caption as: (A) inducing magnet, (D) induced disk, (B) binding-screw for current entering or exiting axis of disc, (B') binding-screw for current entering or exiting circumference of disk, (m) rubber (sliding spring contact) for edge of disk. This was Foucault's and Le Roux's apparatus so this picture was not drawn from Faraday's original machine, but one owned by Faraday's contemporary Leon Foucault. This sketch was downloaded from Google Books. A detailed analysis of the Faraday disc generator is given at http://puhep1.princeton.edu/~mcdonald/examples/magcylinder.pdf. In reading this paper care must be taken to convert the Gaussian units used to the RMKS units before comparing the equations given there to the corresponding equations in this text. Until about 1990 it was very difficult to make a complete three dimensional field calculation of the fields in the Faraday disc generator. A simple one dimensional analysis of the device would tell us that the magnitude of the magnetic induction field normal to the disc and that it was approximately uniform. With the advent of finite and boundary element methods of solving the appropriate partial differential equations subject to the proper boundary conditions such three dimensional problems can be solved on a personal computer. In Appendix F the problem is solved using the finite element solution in particular to find out how good were the assumptions of earlier analysts.

An observer stationary with respect to the disc sees no emf about any path wholly within the disc because both $\mathbf{v}$ and $\partial\mathbf{B}/\partial t$ are zero everywhere on such

Figure 9: Foucault's and Le Roux's copy of Faraday's generator

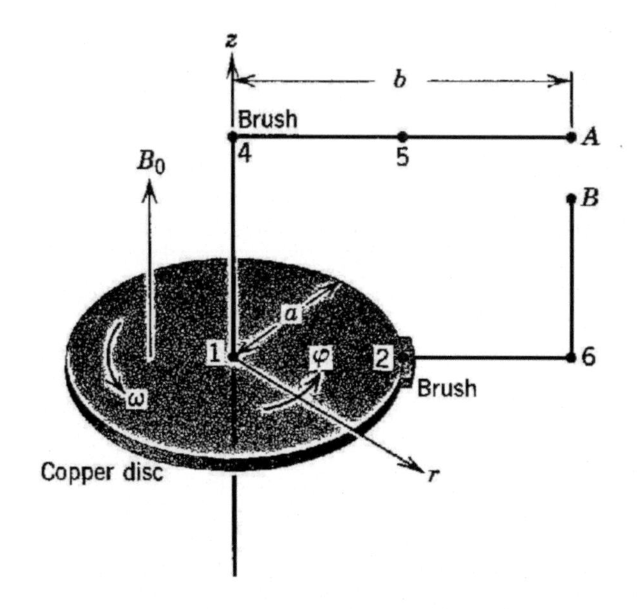

Figure 10: Simplified view of the Faraday disc generator

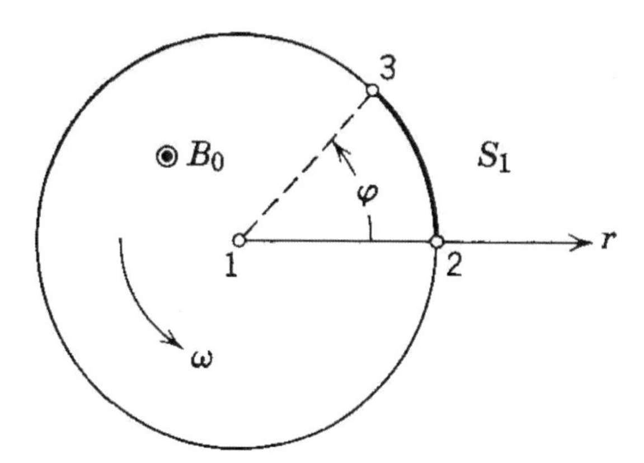

Figure 11: Top view of Fig.10

a path. Therefore no circulating current is present in the disc. If it is assumed no current is drawn from the terminals, the internal IR drop is zero, then the terminal voltage, V_{AB}, is equal to the emf generated in any arbitrary circuit formed by connecting the terminals with a rigid wire.

6.4.1 Observer on brush or external circuit

Referring to Fig.10 and Fig.11 and choosing an observer sitting on either a brush or the external circuit, let us again use Eq.19. To this observer $\partial\mathbf{B}/\partial t = 0$, and since the external circuit is not moving relative to the brush, the electromotive force is given by

$$e = \int_1^2 (\mathbf{v} \times \mathbf{B}) \cdot dl \tag{35}$$

where $\mathbf{v}$ is the velocity of the disc at any point along the path of integration. Because $\mathbf{v} = \varphi r\omega$, $\mathbf{v} \times \mathbf{B} = \mathbf{r}\omega B_0$ and then

$$e = \omega B_0 \int_0^a r\,dr = (\omega B_0 a^2)/2 \tag{36}$$

6.4.2 Observer on disc

Instead of choosing that particular path between the brushes, this same observer could choose a path which is stationary with respect to the disc. In this case the element 1 to 3 of the path is assumed to rotate with the disc and the circuit is completed by the segment 3 to 2 along the edge of the disc. Now the integral of Eq.34 must be evaluated over the path 1 to 3 to 2. Along the path 1 to 3 the result is the same as obtained in Eq.35 (remember the observer is still stationary

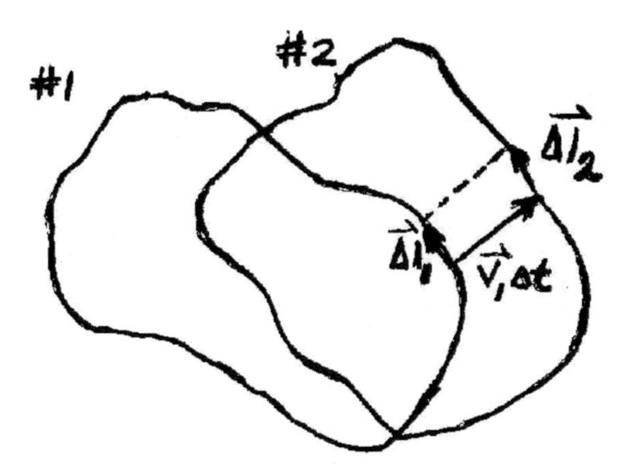

Figure 12: Displaced circuit (no sliding contacts)

with respect to the brushes). Along the segment 3 to 2 $\int_3^2 (\mathbf{v} \times \mathbf{B}) \cdot d\ell = \mathbf{0}$ because $\mathbf{v}$ is parallel to dl. Thus $e = (\omega B_0 a^2)/2$

This illustrates that the relative motion between the observer and the elements of the path is irrelevant. It is only the motion between the elements of the conductor and the observer that is important.

6.5 Simplification of Eq.19

In some cases the expression for emf given by Eq.19 may be written in simpler form. In deriving the new form we consider how the total flux, Φ, through the circuit changes in the small time Δt. For now it is assumed the circuit has no sliding contacts. Let the position of such a circuit at time t be denoted by #1 and its position at time t+Δt by #2 as depicted in Fig.12. Let Δl_1 be an element of the path in circuit #1 which moves to a position Δl_2 in path 2. Let $\mathbf{v}_e$ be the velocity of this element of path Δl_1 Let Φ_1 be the flux through path #1 and Φ_2 be the flux through path Δl_2. Then $\Phi_1 = \int_1 \mathbf{B} \cdot d\mathbf{S}$ and $\Phi_2 = \int_2 \mathbf{B} \cdot d\mathbf{S}$.

The area swept out by Δl_1 in the time Δt is approximately (exactly in the limit) equal to $\mathbf{v}_e \Delta t \Delta l_1$. Thus Φ_2 can be rewritten as

$$\Phi_2 = \int_1 \mathbf{B}(t + \Delta t) \cdot dS + \int_1 \mathbf{B}(t + \alpha \, \Delta \, t) \cdot \mathbf{v}_e \, \Delta \, t \Delta l_1 \text{for } 0 \leq \alpha \leq l \qquad (37)$$

The total rate of change of flux is given by

$$d\Phi/dt = \lim_{\Delta t \to 0} (\Phi_2 - \Phi_1)/ \, \Delta \, t \qquad (38)$$

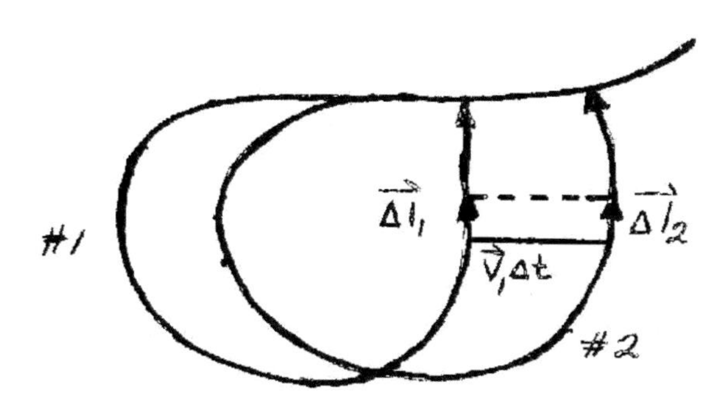

Figure 13: Displaced circuit (with sliding contacts)

which upon substitution of various terms becomes

$$d\Phi/dt = \lim_{\Delta t \to 0}\{[\mathbf{B}(t+\Delta t) - \mathbf{B}(t)] \cdot d\mathbf{S}\} + \lim_{\Delta t \to 0} \oint_1 \mathbf{B}(t + \alpha \bigtriangleup t) \cdot \mathbf{v}_e \times dl_1 \quad (39)$$

becoming in the limit as $\bigtriangleup t \to 0$

$$d\Phi/dt = \int_1 \partial\mathbf{B}/\partial \cdot d\mathbf{S} + \oint \mathbf{B} \cdot \mathbf{v}_e \times dl_1 \quad (40)$$

However, $\mathbf{B} \cdot \mathbf{v}_e \times dl_1 = \mathbf{B} \times \mathbf{v}_e \cdot dl_1 = -\mathbf{v}_e \times \mathbf{B}{\cdot}dl_1$.Therefore it follows that the total time rate of change of flux is given by

$$d\Phi/dt = -\int_S \partial\mathbf{B}/\partial\mathrm{t} \cdot \mathbf{dS} + \oint \mathbf{v}_e{\times}\mathbf{B}{\cdot}dl_1 \quad (41)$$

Eq.41 is very similar to Eq.19, the only difference being that it contains $\mathbf{v}_e$ which is the velocity of the path element dl, whilst the other contains velocity $\mathbf{v}$ of the conductor at the element dl. This same result is obtained if there are sliding contacts in the circuit that slide in such a manner that the change in length of the circuit due to this sliding contact is a continuous function of time. If a sliding brush is to remain in contact with the circuit under these conditions, the relative velocity between the contact and the circuit must be directed along the tangent to the path at the point of contact. Fig.13 shows two consecutive positions of the circuit from the point of view of an observer stationary with respect to the circuit at the point of contact. It is clear that the same analysis can be applied as before.

As we have shown in Example 4, in the calculation of emf it is sometimes possible to choose a path of integration such that the velocity of every path element is the same as that of the conductor at that point(e.g. the path moves along with the conductor). If and only if such a path is chosen the expressions for the electromotive force and the rate of change of flux become identical and hence, $e = -d\Phi/dt$. As a matter of fact, such a path may always be found and this formula applied it the circuit consists of a single, continuous conductor or if it contains sliding contacts which move such that the path length between any two points fixed on the conductors is a continuous function of time. It is interesting to note that the evaluation of this result is completely independent of the observer so that his position need not be specified when the emf is computed in this manner.

6.6 Reworking previous examples with Faraday's Law

Here we shall reconsider previous examples and show how they can be worked using Faraday's Law as expressed in Eq.8. Example 6.1 is reworked as follows: Let the distance from the left edge of the loop to the slider be x. Then the flux in the loop is given by

$$\Phi = \int_S \mathbf{B} \cdot d\mathbf{S} = (\mu_0 I x/2\pi) \int_a^b r^{-1} dr = (\mu_0 I x/2\pi) \ln(b/a) \tag{42}$$

Thus

$$e = -d\Phi/dt = -[(\mu_0 I/2\pi) dx/dt] \ln(b/a) = -(\mu_0 I v/2\pi) \ln(b/a) \tag{43}$$

which is the same result obtained in example 1.

For Example 6.2 the total flux in the rotating coil is $\Phi = \int_S \mathbf{B} \cdot d\mathbf{S} = 2rlB_0 \sin(\theta)$ and taking the derivative of the total flux with respect to time yields

$$e = -d\Phi/dt = -2rlB_0[\cos(\theta)] d\theta/dt = -2rlB_0 \omega \cos(\omega t) \tag{44}$$

which is the same result obtained in example 6.2

In Example 6.3, $\Phi = 2rlB_0 \sin(\theta) = 2rlB_m \cos(\omega t) \sin(\omega t) = rlB_m \sin(2\omega t)$. Taking derivative of the total flux with respect to time yields

$$e = -2rl\omega B_m \cos(2\omega t) \tag{45}$$

which is the same result obtained earlier.

The configuration of Example 6.4 needs somewhat more careful consideration because it has a distributed conductor instead of a thin wire as was the case in the previous examples. This is the simple Faraday disc generator that many claim cannot be explained by Faraday's Law. The flux Φ is the instantaneous flux crossing the surface bounded by a path which moves with the conductor.

The geometry is shown in Fig.13. Because the leads to the brushes have been drawn in a plane parallel to $\mathbf{B}_0$, the only flux linking the path is that through the area, S, in the right sketch denoted as ABC. In that case $\Phi = \int_S \mathbf{B} \cdot d\mathbf{S} = \int_0^{\omega t} \int_0^a B_0 r dr d\theta = (B_0 a^2 \omega t)/2$. Therefore $V_{AB} = e = -(B_0 a^2 \omega)/2$ and

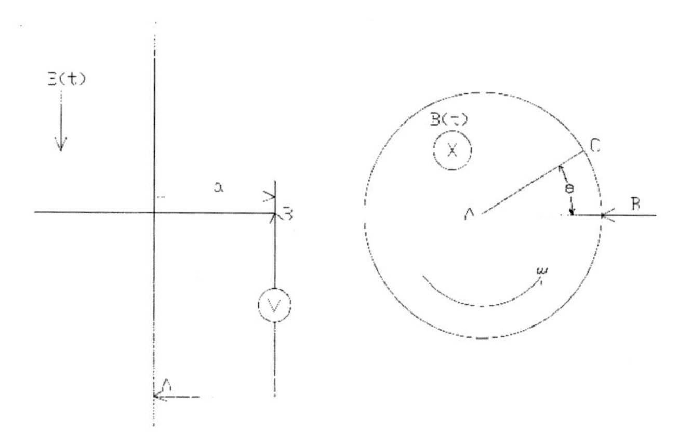

Figure 14: Faraday disc generator

is the same result previously obtained. Even if the wires from the brushes had not been in a plane parallel to $\mathbf{B}_0$, the flux through the circuit would not change with time so that the resultant emf would be the same as found above.

6.7 Faraday disc generator in an alternating magnetic field

An interesting situation arises when the magnetic field in Examples 6.4 and 6.5 is sinusoidal in time rather than constant. This problem has resulted in a somewhat heated controversy in many journals. It should lead to no confusion if it is approached carefully, however. Because the conducting disc is located in a changing magnetic field, there will be eddy currents flowing so as the reduce the changing magnetic field in the disc. They will flow azimuthally and be small enough to neglect any change they make in the axial magnetic field if the disc is thin or it has large resistivity. The axial field shall be assumed not to be distorted by these eddy currents. The following method of analysis is correct even if this assumption is not valid, but in that case the field problem must be solved to find the distribution of the magnetic induction, $\mathbf{B}$ in the disc. To illustrate this Fig.15 is presented here. In the lower left corner of this figure is a good conducting cylindrical disc immersed in the uniform field from the AC magnet in the right corner. Between them is an air gap one tenth the width of the left conductor. A finite element solution of Maxwell's equations resulted in the magnetic induction flux cylinders shown in this figure. This solution is described in detail in Appendix A. Although the variation in the magnetic induction could easily be included in an emf calculation, the field calculation is more difficult. The skin depth in that conductor is about 0.4 causing distortion of the flux tubes to the right of the air gap. In Fig.13 nothing is changed except the magnetic induction that is now given by $\mathbf{B}_0 = B_m \cos(\beta t)$. First it is solved by Eq.19 and taking the path AB and assuming the wires to the brushes lie in a plane parallel to $\mathbf{B}$ which is a necessary assumption. Then the first term is

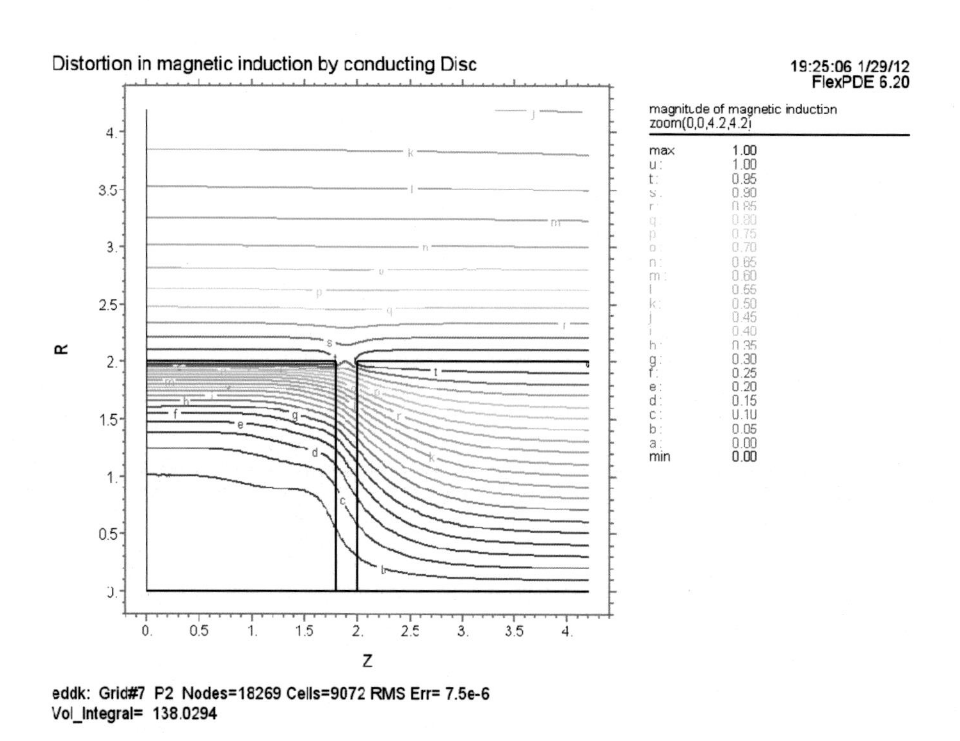

Figure 15: Eddy current distortion of a uniform AC magnetic field

zero because everywhere $\mathbf{dS}$ is perpendicular to $\partial\mathbf{B}/\partial t$. The second term is the same as in the first part of Example 4 and

$$e_1 = -(1/2)B_0\omega a^2 = -(1/2)B_m\omega a^2\cos(\beta t) \tag{46}$$

Solving the same problem using the path ACB, the first term in Eq.19 is not zero. Here the first term is given by

$$-\int_S \partial\mathbf{B}/\partial t\cdot\mathbf{dS} = \int_0^{\omega t}\int_a^a B_m\beta\sin(\beta t)r\,dr\,d\theta = (B_m a^2\omega/2)\beta t\sin(\beta t) \tag{47}$$

and

$$\oint \mathbf{v}\times\mathbf{B}\cdot dl = B_0 a^2\omega/2 = (B_m a^2\omega/2)\cos(\beta t) \tag{48}$$

The emf is given by the addition of Eq.47 and Eq.48 which yields

$$e_2 = (B_m a^2\omega/2)[\beta t\sin(\beta t) - \cos(\beta t)] \tag{49}$$

Next we calculate the emf from $e = -d\Phi/dt$.Here $\Phi = (B_m a^2\omega t/2)\cos(\beta t)$ and executing and negating the indicated time derivative yields

$$e_3 = (B_m a^2\omega/2)[\beta t\sin(\beta t) - \cos(\beta t)] \tag{50}$$

The calculations of the last two methods taken on path ACB agree exactly whereas the calculation based upon path AB yields a different result giving rise to arguments as to which is correct. However, there is no reason why they should give the same result. There is no principle that states the emfs about different paths must be the same. In fact, Kirchhoff's voltage law assures us that the emf about different paths is not the same. These results will disturb those who confuse emf with terminal voltage which must be independent of the means of its calculation. In the previous examples the emf was equal to the terminal voltage provided the voltmeter did not draw any current; in this example eddy currents cause voltage drops that are not negligible. For example consider the simple DC circuit shown below in Fig.16, for which there is surely no argument:Here the total emf in loop ACDBA is 2 volts whilst in loop ACEFDBA it is 5 volts. The quantity that must be the same no matter who calculates it is the terminal voltage, not the emf. It has been shown earlier that in general the terminal voltage V_{AB} equals the emf minus the internal IR drop. In all of the other examples the current was assumed to be zero, but in the case of the Faraday disc rotating in an alternating magnetic field there are eddy currents whose voltage drops cannot be neglected even though the distortion of the magnetic field is negligible. By symmetry, the eddy currents flow in concentric circles about the z axis; the observer who chooses the radial path AB finds that the eddy current J_φ is perpendicular to his path. Here $V_{AB} = e_3 - \int_A^B (J_\varphi/\sigma)\boldsymbol{\varphi}\cdot\mathbf{r}dr$. For observers whose paths of integration include segment CB the second term

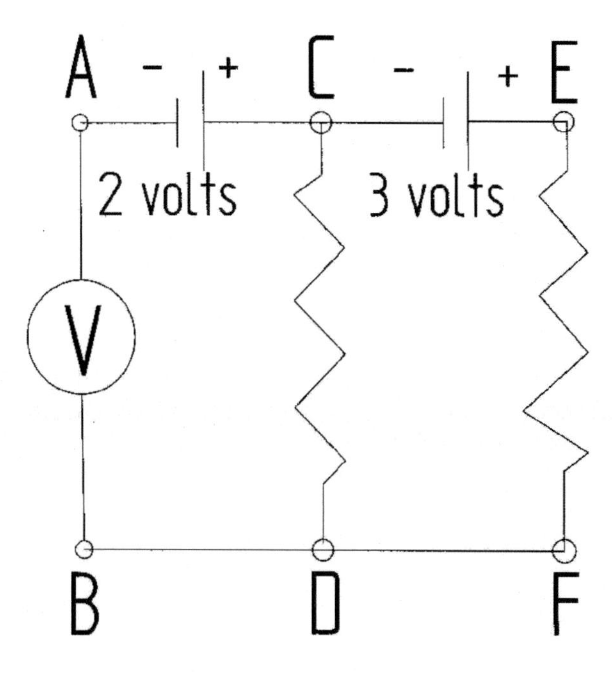

Figure 16: Simple DC circuit

of the expression for the terminal voltage is not zero. Because E_φ does not vary with φ, $E_\varphi = -(r\beta B_m/2)\sin(\beta t)$ and thus

$$\int_C^B \mathbf{E}\cdot dl = (B_m a^2 \omega/2)\beta t \sin(\beta t) \tag{51}$$

and thus for those observers

$$\mathrm{V}_{AB} = e3 - \int_C^B \mathbf{E}\cdot dl = (B_m a^2 \omega t/2)\cos(\beta t) \tag{52}$$

and *all three observers agree on the terminal voltage.*

If the disc has a radial slit or some other discontinuity, the eddy currents would not be angular and the analysis would be more complicated although the methods derived herein are perfectly general.

6.8 Faraday disc in a nonuniform magnetic field

Unless great care is taken the magnetic induction in the air gap of real magnets is likely to be nonuniform due to fringing and saturation of the iron in the core corners. In such cases $\oint \mathbf{v} \times \mathbf{B}\cdot dl$ must be done numerically from the results of a complete calculation of the magnetic induction. Here a cylindrical core excited by a permanent magnet is considered. In Fig.17 is shown the geometry. This magnet is shown in cylindrical coordinates and is the same for all angles. The top and left side of the magnet is made of soft high permeability iron and the bottom region on the right side is the permanent magnet driving the flux. At the left in the region where $0 \prec r \prec 20$ is the air gap in which the Faraday disc rotated about the $r = 0$ axis with angular velocity ω .Fig.18 exhibits the variation of B_z in the air gap and Fig.19 yields the product rB_z. In Fig.19 it is clear that most of the emf is generated for $1 \le r \le 9$ distance units even though the air gap is 10 units wide. (This calculation was done in cgs units because there $\mu_0 = 1$ which is very convenient.) B_z is almost constant in the interval $1 \le r \le 8$. The values of $\int_1^r rB_z dr$ are given in Fig.20. Although the effect of fringing becomes apparent at $r \cong 8$, the emf can be enlarged by increasing the disc outer radius to 18.4 distance units. In summary, this example shows how the complete field solution can be used to find the emf. The solution presented here neglects the nonlinear characteristics and the inherent anisotropy of the iron. Although these can be handled in the solution to the field problem they are not in the scope of this text.

6.9 Confusion due to flux linkages changing with time

Another source of confusion is the belief that there is an emf generated when ever the flux linkages of the circuit change with time. For example, is there

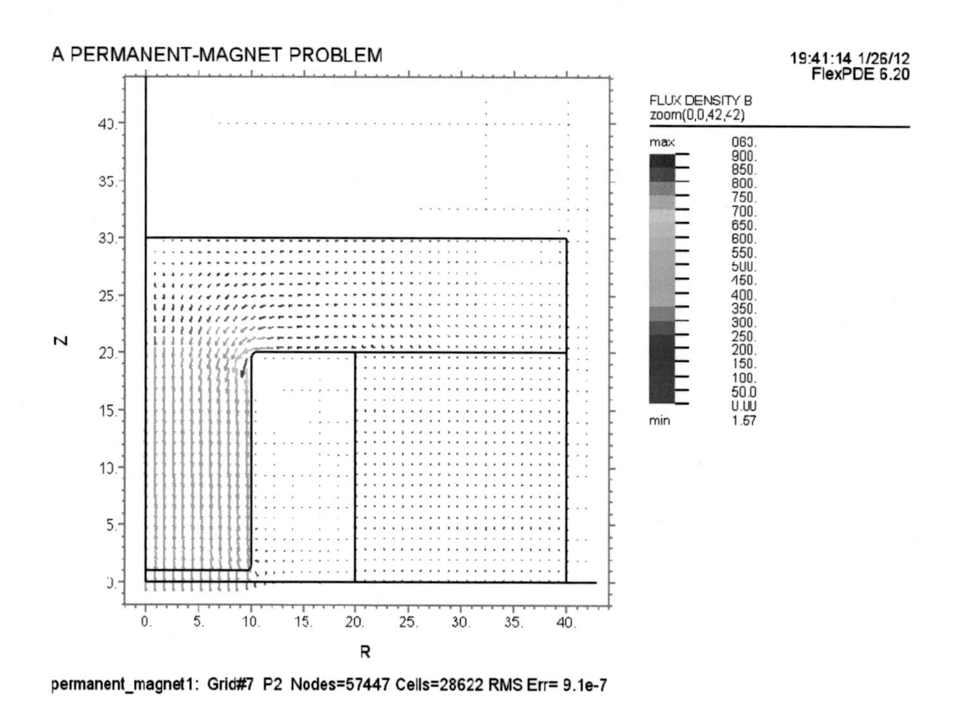

Figure 17: Cylindrical permanent magnet

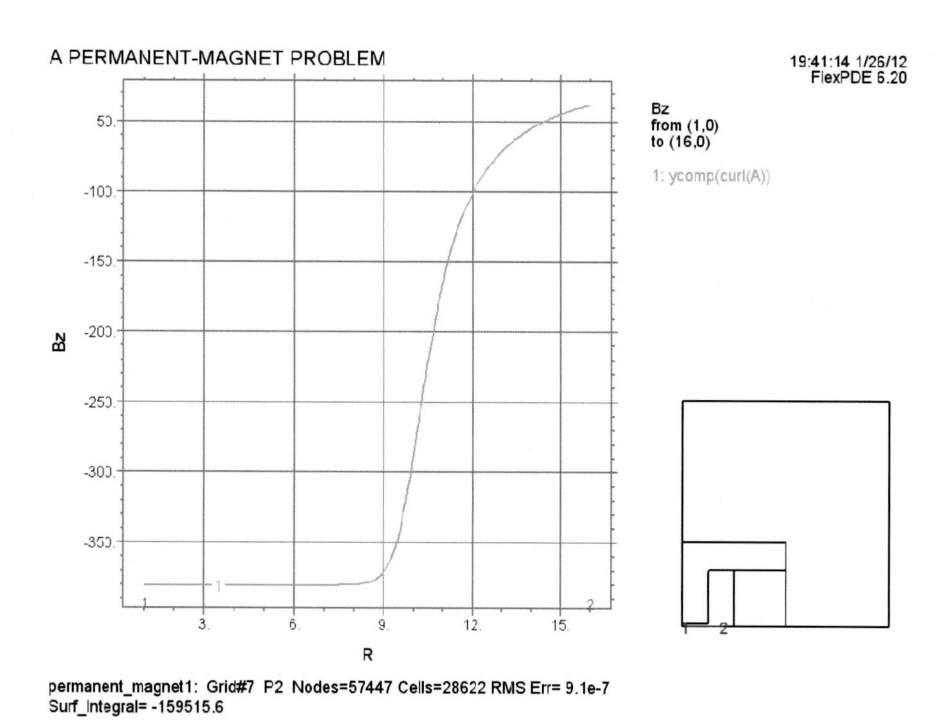

Figure 18: B_z in the air gap

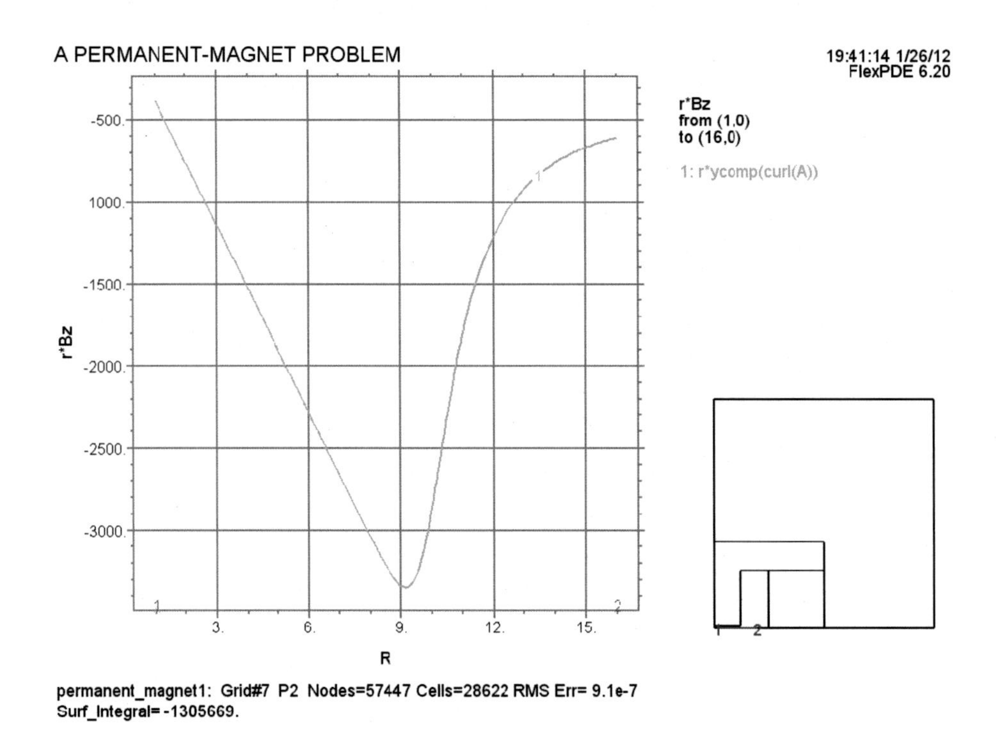

Figure 19: The product of B_z and r in the air gap

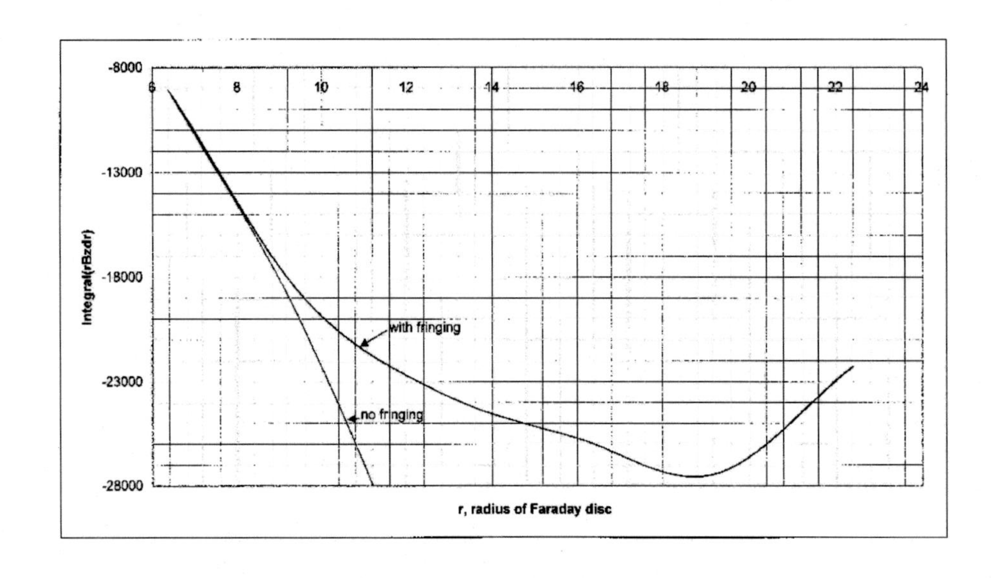

Figure 20: The effect of fringing on emf

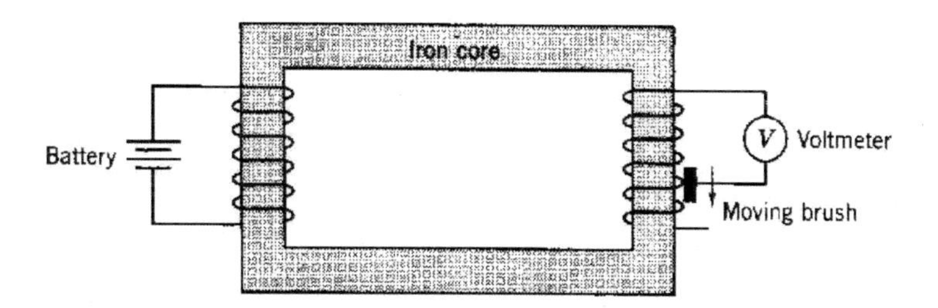

Figure 21: Time varying flux linkage

an emf generated when the slider in Fig.21 is moving?The flux linked by the circuit is certainly changing with time. However, this temporal rate of change of flux is due neither to a motion of a conductor through a magnetic field nor to a changing magnetic field. Hence, by $e = -\int_S \partial\mathbf{B}/\partial\mathrm{t} \cdot \mathbf{dS} + \oint \mathrm{v} \times \mathbf{B} \cdot dl$, the emf is zero. Faradays Law cannot be applied because the path length is not a continuous function of time, as was mentioned previously.

6.10 Faraday disc revisited

For another short example, we shall again consider the Faraday disc generator but in the form shown in Fig.22. The discand magnet are circular cylindrical shape.

(a) Would the voltmeter read if the disc were revolving (clockwise looking downward) and the voltmeter and field magnetic circuit stationary with respect to the earth? Yes this is the same problem considered in Example 6.4 where the emf was shown to be equal to $(1/2)B_0\omega a^2$.

(b) Would the voltmeter read if the voltmeter and its connection and the brushes were revolving (clockwise looking down) and the disc and field poles were stationary? Yes, because the lead to the lower brush is moving through a magnetic field resulting in an emf of $-(1/2)B_0\omega a^2$.

(c) Would the voltmeter read if the field were revolving (clockwise looking down) and the disc and the voltmeter were stationary? This might be accomplished by letting the magnetic circuit rotate. N0, since to an observer stationary with respect to the disc and the voltmeter, there is no velocity of any segment of the circuit and no varying magnetic field.

6.11 A homopolar generator

As stated earlier, it is often very much easier to calculate the emf in some one particular manner than in any other way, although it is impossible to say that

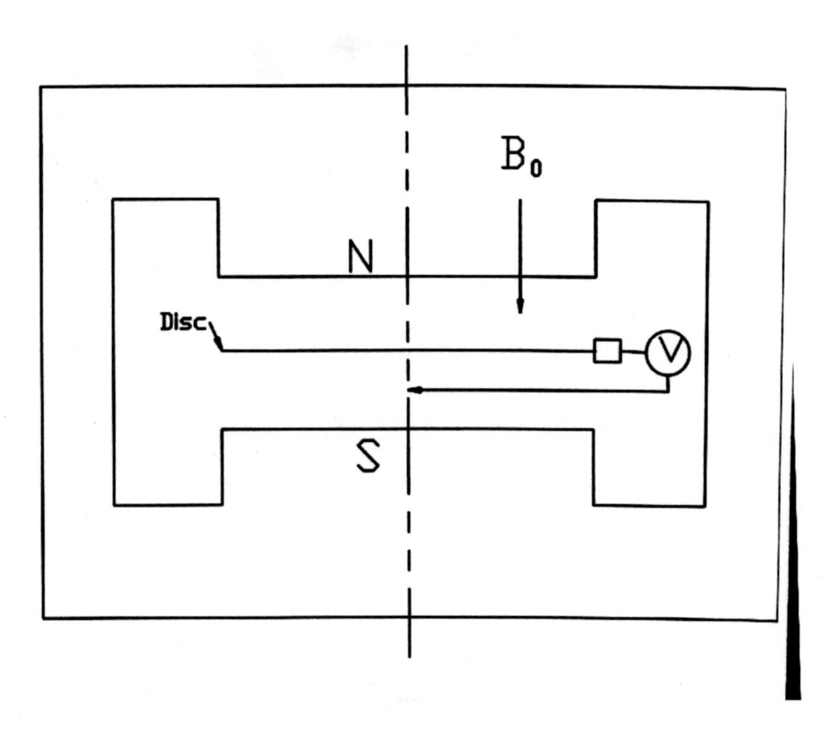

Figure 22: Permanent magnet excited homopolar generator

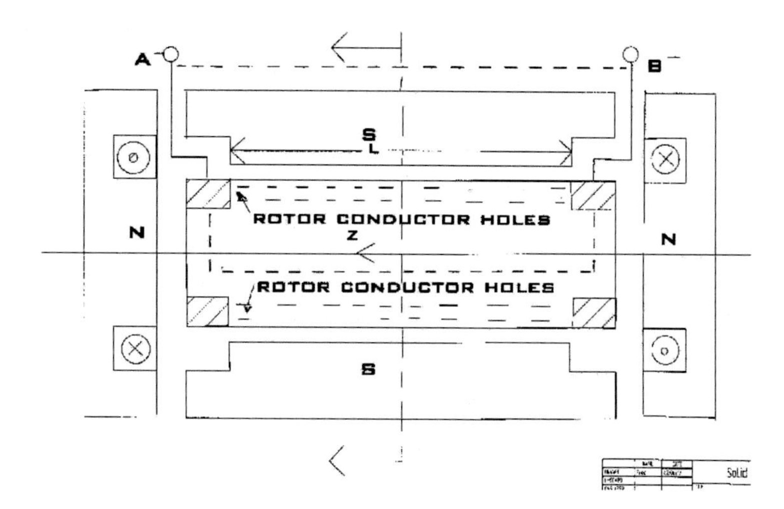

Figure 23: A Homopolar generator with iron parts

any one particular method will be the simplest for all problems. In this example the emf will be calculated in four different ways to illustrate this difference in complexity and also to show how to treat the case in which the conductors are imbedded in iron as in squirrel cage induction motors, etc.

Consider a cylindrical homopolar generator whose armature conductors are imbedded in the iron rotor as shown in Fig.23 and whose stator is divided into three sections along the axis. It is assumed that the air gap flux density is radial and independent of both θ and z.

6.11.1 Faraday's Law

The simplest method of calculating the emf in this case is to use the simple form of Faradays Law, given in Eq.8. Here the flux passes through the armature of length L and rotor radius a. Then $\phi = B_0 La\theta$ where θ is the angle curling around the z axis. Hence the terminal voltage is given by $V_{AB} = e = -B_0 Lad\theta/dt = -B_0 La\omega$ where ω is the angular velocity of the rotor.

6.11.2 Observer on the rotor using Eq.19

Next we shall use Eq.19 and place the observer on the rotor. To this observer there is no changing field anywhere in space so that the first term is zero. To him the armature conductors are stationary so that the only conductors he sees moving through a magnetic induction field are the leads from the brushes there they pass through the air gap in the stator. This is illustrated in Fig.24. The line integral will be the same over each wire including the one on the left side by symmetry.

Then $e = 2 \int_{r=b}^{r=c} \mathbf{B} \times \mathbf{V} \cdot d\ell$. Here $V = r\omega$ and is directed into the paper and

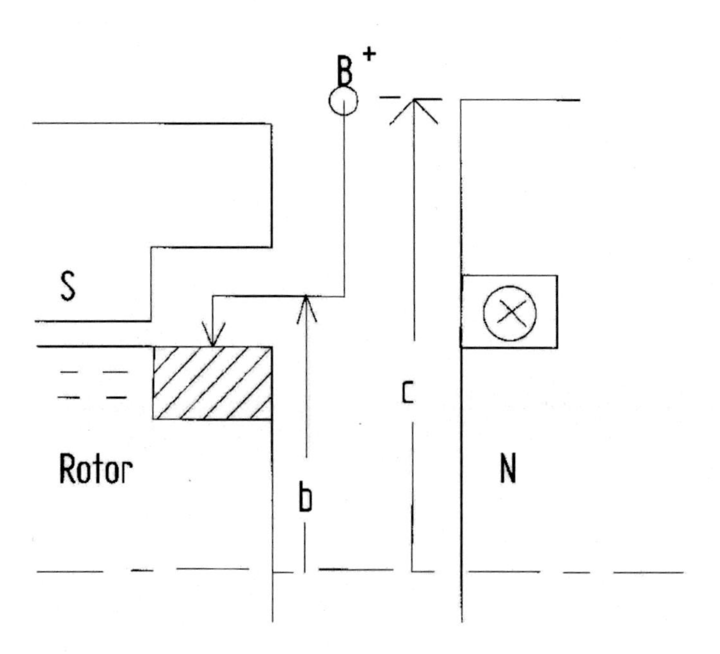

Figure 24: Brush and conductor on the right side of Fig. 22

$d\ell = d\mathbf{r}$. In vector notation: $\mathbf{V} = r\omega\boldsymbol{\theta}, d\ell = \mathbf{r}dr$ and $\mathbf{B} = B_r\mathbf{r} + B_z\mathbf{z}$. Then $\mathbf{B} \times \mathbf{V} \cdot d\ell = r\boldsymbol{\omega}(B_z\mathbf{r} - B_r\mathbf{z}) \cdot r\,dr = B_z\boldsymbol{\omega} r\,dr$ whence $e = 2\omega \int_b^c B_z r\,dr$. However $2\pi \int_b^c -B_z r\,dr = \phi_t/2$ where ϕ_t is the total flux which equals $2\pi a L B_0$. Then the result becomes

$$e = -2\omega\phi_t/4\pi = -\omega\phi_t/2\pi = -(\omega/2\pi)(2\pi a L B_0) = -B_0 L a\omega \tag{53}$$

which is the same result obtained from Faraday's Law.

6.11.3 Observer on the stator

A third method for calculating the emf is to place the observer on the stator. If it could be assumed that the conductors are on the surface of the rotor with no slots, the analysis would be very simple because $\partial\mathbf{B}/\partial\mathbf{t}$ would be zero everywhere and $\int \mathbf{v} \times \mathbf{B} \cdot dl$ over the armature conductors would be $-B_0 L a\omega$ checking with the previous results. However in the case being considered, the conductors are actually imbedded in the iron and since iron has a much higher permeability than copper, there shall be a very small magnetic induction in the conductors. In Fig.25 is a plot of the magnetic induction vectors in the vicinity of a conductor hole in the rotor of the homopolar machine depicted in Fig.23. The vectors in the hole are too small to see. The geometry of the first quadrant of this generator is shown in Fig.26. Note the conductor holes every thirty degrees. Along white arcs **B** normal to the arcs is calculated. These results are given in the graphs of Fig.27 and Fig.28.

In the plots of these figures we see how very small the magnetic induction is in the holes. The magnetic permeability was set to $\mu = 1000\mu_0$ and it might will be larger as relative permeabilities as large as 10^6 are obtainable with the addition of nickel and special heat treating. In Fig.28 there are some variations of the radial magnetic induction probably caused by small inaccuracies in the finite element calculation and exaggerated by the scale on the graph. Here B_r is essentially constant with a value around 0.5945.

Fig.27 shows that $\int \mathbf{v} \times \mathbf{B} \cdot dl$ taken over the conductors is very much less than it would be in the aforementioned hypothetical case. This does not mean that this method gives the wrong answer because an observer on the stator see a time rate of change of **B** at a point in the rotor along the same radius as the center of the conductors. As the conductors move by he will see the field alternating between small and large as they go by. Thus this observer must evaluate both of the integrals in Eq.19. Choosing the path shown below, the second integral becomes

$$\int \mathbf{v} \times \mathbf{B} \cdot dl = -\int_{-L/2}^{L/2} \omega h B_r(h, \theta_0, z)dz \tag{54}$$

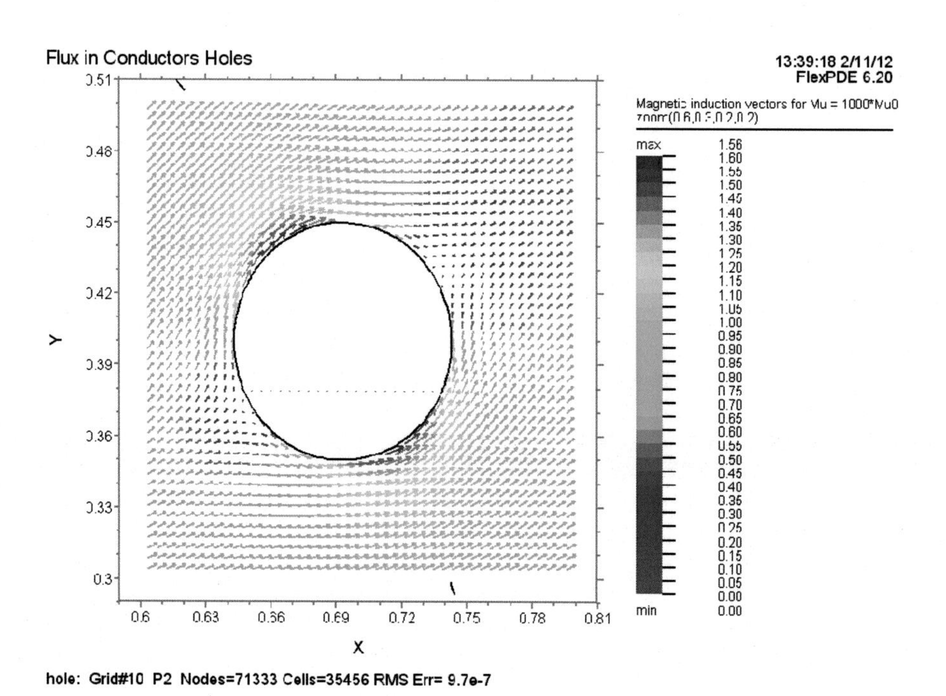

Figure 25: Magnetic induction vectors in vicinity of hole

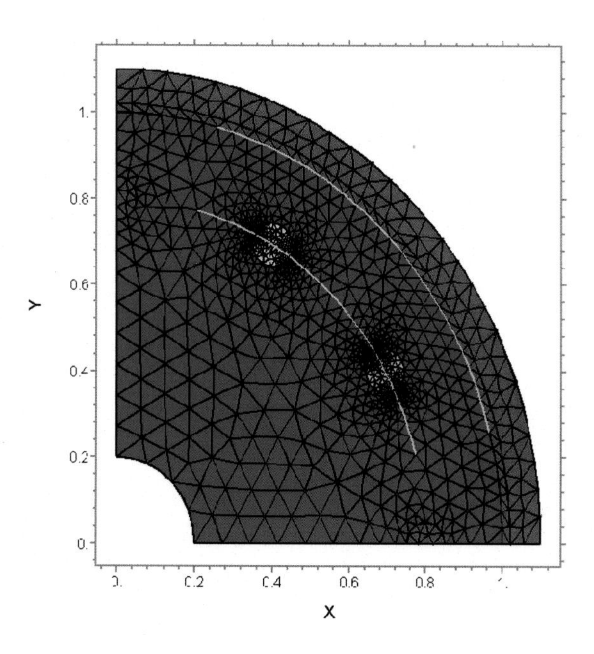

Figure 26: A quadrant of the homopolar generator

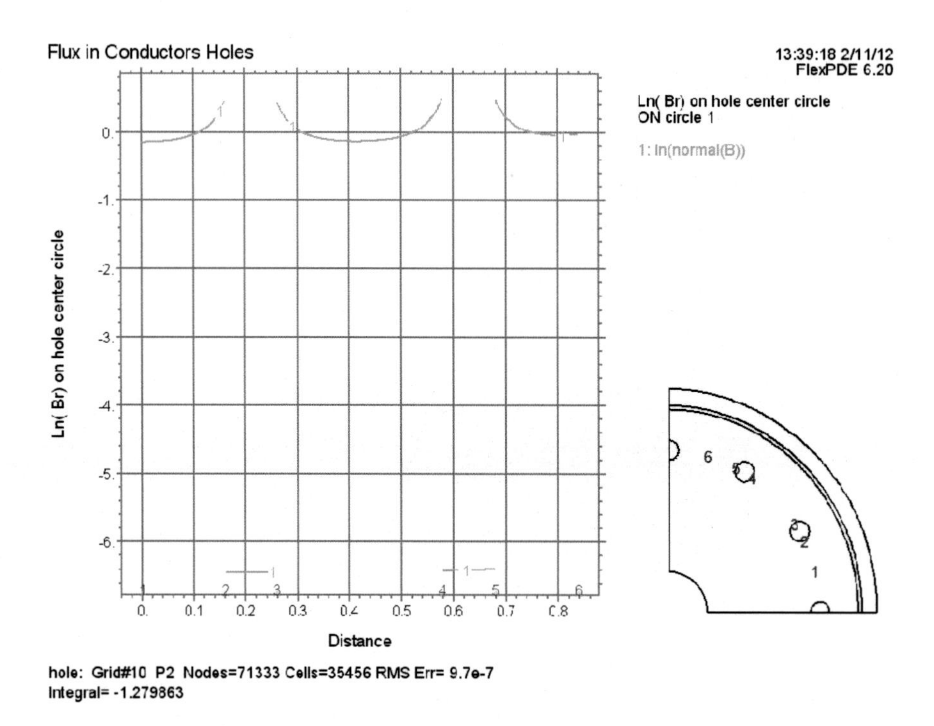

Figure 27: Natural log of B_r through the conductor holes

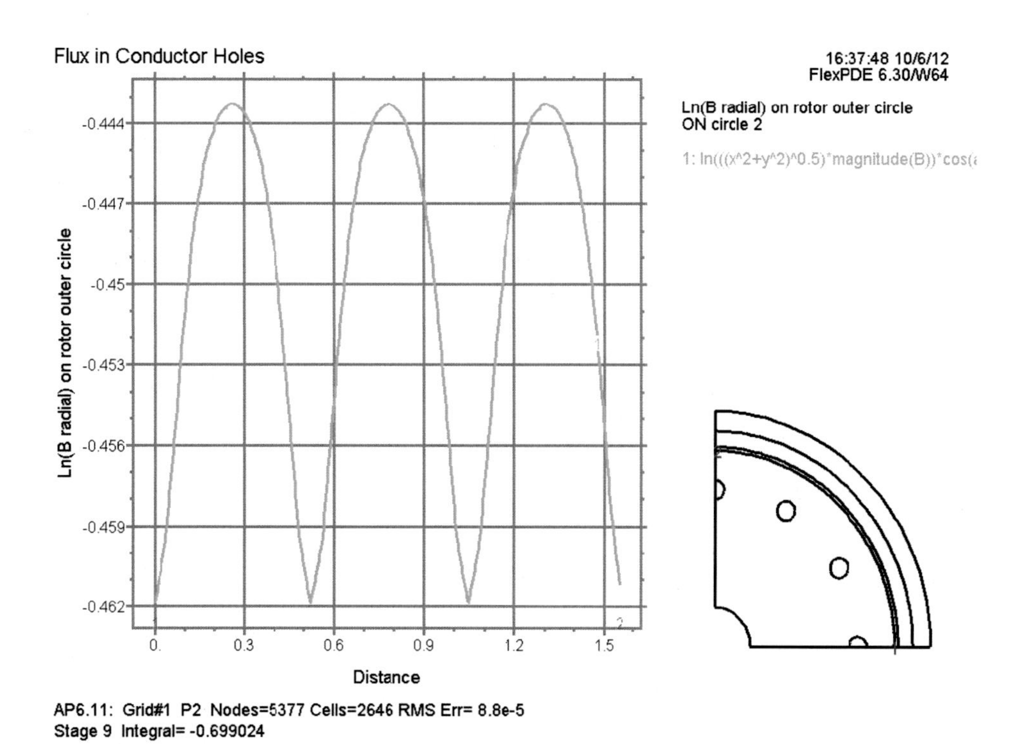

Figure 28: Natural log of B_r on rotor outer circle

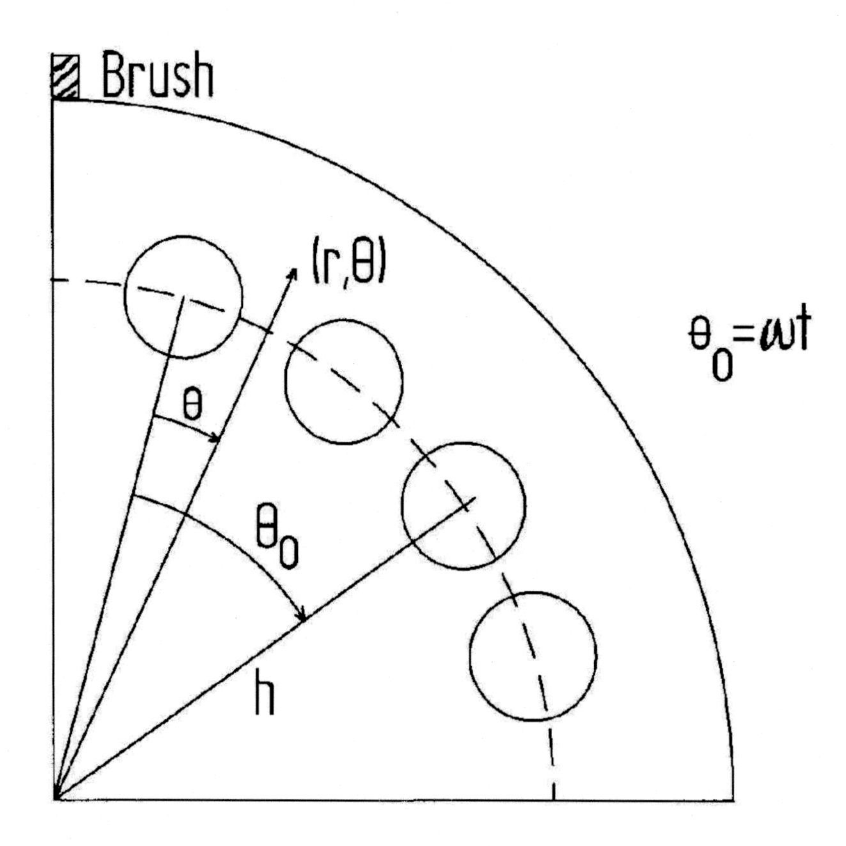

Figure 29: Rotor seen by observer on stator

The negative of the first term is given by

$$\int_S \partial\mathbf{B}/\partial\mathbf{t}\cdot\mathbf{dS} = \int_{-L/2}^{L/2} d/dt\left\{h\int_0^{\theta_0}\partial B_r(h,\theta,z)/\partial t d\theta + \int_h^a[\partial B_\theta(r,0,z)/\partial t]dr\right\} \tag{55}$$

By the rules for differentiating under the integral we obtain

$$h\int_0^{\theta_0}\partial B_r(h,\theta,z)/\partial t d\theta = d/dt\int_0^{\theta_0}B_r(h,\theta,z)hd\theta - B_r(h,\theta_0,z)hd\theta_0/dt \tag{56}$$

and

$$\int_h^a[\partial B_\theta(r,0,z)/\partial t]dr = [d/dt\int_h^a B_\theta(h,\theta,z)dr] \tag{57}$$

Combining the results of Eq.55 and Eq.56 and substituting them into Eq.54 yields

$$\int_S \partial\mathbf{B}/\partial\mathbf{t}\cdot\mathbf{dS} = \int_{-L/2}^{L/2}\left[\begin{array}{c} d/dt\left\{\int_0^{\theta_0}B_r(h,\theta,z)hd\theta + \int_h^a B_\theta(h,\theta,z)dr\right\} \\ -B_r(h,\theta_0,z)h\omega\end{array}\right]dz \tag{58}$$

Here by the continuity of magnetic flux there results

$$\int_0^{\theta_0}B_r(h,\theta,z)hd\theta = \int_0^{\theta_0}B_r(a,\theta,z)ad\theta = B_0a\theta_0 \tag{59}$$

and because $\int_h^a B_\theta(h,\theta,z)dr$ is not a function of time

$$-\int_S \partial\mathbf{B}/\partial\mathbf{t}\cdot\mathbf{dS} = -B_0aL\omega + \int_{L/2}^{L/2}B_r(h,\theta_0,z)h\omega dz \tag{60}$$

Adding the result of Eq.60 to Eq.54 yields the emf (noting that the integrals over dz cancel) which is $e = -B_0aL\omega$ being the same as the previous observers obtained. This observer on the stator might have chosen a path along the surface of the rotor to calculate the emf. This is included in examples for the interested reader to consider. Another variation of this example would be to consider a more practical construction of this homopolar generator where the stator is a single piece of iron and the leads from the brushes pass through small holes in this stator instead of through an annular gap. It is obvious that Faraday's Law yields the same emf for this case as it did for the three piece stator. It is left to the ambitious reader to show that the emf as calculated by an observer on the rotor is also the same, by actually carrying out the evaluation from this viewpoint.

6.11.4 Homopolar generator summary

All of this involved elementary mathematics serves to show once again that the answer must be the same no matter how the problem is approached and it also demonstrates that sometimes one method may be much easier to complete than another.

6.12 A common direct current generator

As a sequel to the previous example in which the conductors were imbedded in the iron, we shall now consider the common DC generator in which the armature conductors lie in slots about the periphery of the rotor. In Fig.30 is shown a pair of DC generators excited by a permanent magnet. The left rotor has no slot whilst the right has a slot. In the center at the top is the permanent magnet. At each side is the stator iron. In the lower left and right are one quadrant of the rotors. The small green box on the right rotor is the slot that along with the air gap above the rotor is air. We shall compare the fields in both rotors having identical excitation. Many texts compute the emf for such a machine by using the formula $e = BLv$ (apparently taking the stator as the reference). The value of B used is the magnetic induction or flux density in the air gap. Strictly speaking this approach is incorrect because the flux density in the slot where the conductor would reside is almost zero. This was the case in Example 9.

In Fig.31 the flux tubes in this configuration are shown. Notice the flux tubes that are avoiding the slot in the right rotor. The observer on the stator will also see a changing field as the teeth move by so that $\int_S \partial \mathbf{B}/\partial t \cdot \mathbf{dS} \neq \mathbf{0}$ and hence must be included with the BLv term. Nevertheless this simple textbook treatment does give the correct answer and hence should not be too strongly condemned; rather we shall try to justify its use and show why it yields the correct result.

Consider the one turn coil shown in the lower right corner of Fig. 30. This figure shows the second quadrant of the right rotor that has a slot for conductors. The emf of this coil may be found by applying Faraday's Law over the area embraced by the coil. This area is bounded by a radial line passing through the center of the slot to the center of the rotor and through the rotor and the corresponding slot in the third quadrant. It is obvious that, neglecting the small amount of flux leading across the slot, at the time this coil is perpendicular to the pole axis all the air gap flux will cross this area. When the coil has rotated one-half revolution, all of the flux will link it in the opposite direction. See Fig.33 and Fig. 34 that show the magnetic induction vectors in the slot are very small in contrast to the same location in the left rotor as depicted in Fig.35. To obtain the field solution given here, a relative permeability of 10,000 is used for the iron which is treated as linear. It is noteworthy that for relative permeabilities greater than about 10 the slot magnetic induction is greatly reduced. Note that the generator on the left is the same as the one on the right except its conductor is wound on the periphery of the smooth rotor. The flux through this coil varies between the same limits in the same time interval as in the original machine.

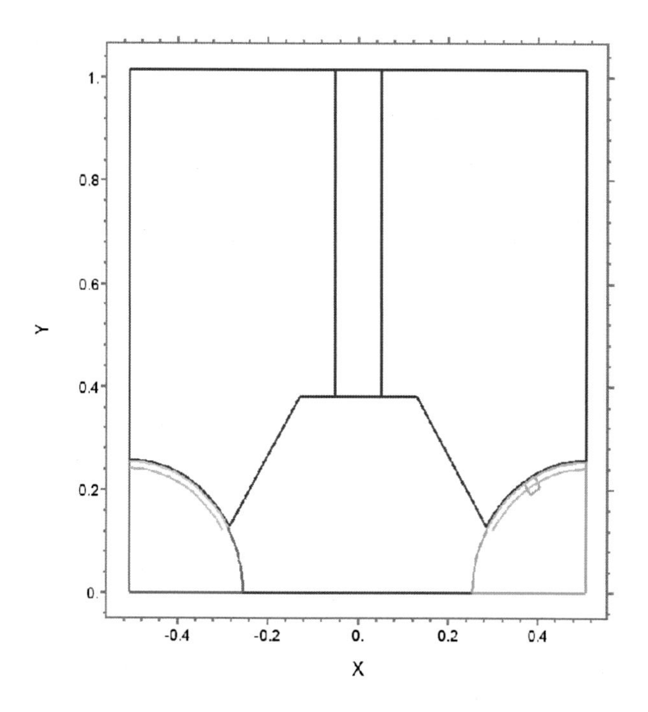

Figure 30: A pair of direct current generators

Fig.36 and Fig.37 exhibit the normal or radial component of magnetic induction, B_r along the periphery of the slotted rotor and the smooth rotor. B_r is almost constant on the smooth rotor whereas on the slotted rotor it is almost constant until it nears the slot. At that station it increases suddenly before diminishing to almost nothing at the slot. The sudden increase is due to the flux crowding seen in Fig.31. B_r is shown along the paths in Fig.32 and Fig.33 that are 1/2" below the rotor surfaces. Clearly the field in the middle of the slot is very small. The larger the permeability of the iron the smaller the slot field becomes. In these calculations $\mu_{iron} = 10000\mu_0$. Even a relative permeability of 10 causes the slot field to be small. The emf in the coil of the right rotor may be found by applying Faraday's Law over the area bounded by the coil. The small amount of flux going radially across the slot is neglected with respect to the rest of the radial flux. Then is obvious that at the time this coil is perpendicular to the pole axis all of the air gap flux will cross this area. When the coil is rotated one-half revolution, the same flux will link it in the opposite direction. The left machine is identical to the right except it has no slot and the coil must be placed on its surface. The flux varies through the same limits in the same time interval as in the machine on the right. The following equations show that the average emf is the same for both machines. Let T be the time it takes for one-half revolution, t =0 when the coil is perpendicular to the pole axis and ϕ_t the total air gap flux under each pole. Then

$$e_{ave} = (1/T)\int_0^T edt = (1/T)\int_0^T (-d\phi/dt)dt \tag{61}$$

$$e_{ave} = (1/T)\int_{-\phi_t}^{\phi_t} d\phi = 2\phi_t/T \tag{62}$$

Actually the instantaneous emf will not be identical in both machines because of the distortion of the field as the slot moves under the pole tip. In effect the substitution of the left generator for the right neglects the slot harmonics created in the right rotor. A more exact calculation could be made using the aforementioned field calculation results by Eq.19 or by Faraday's Law which would be the simpler. This would be a good problem for the ambitious reader.

Here the BLv formula is correct more or less by luck. There are numerous other cases where it yields completely erroneous answers.

6.12.1 Complete direct current generator

This one turn analysis can be extended to the complete DC generator by noting that the total voltage between the brushes is just the sum of the instantaneous voltages in each coil and hence is equal to the number of conductors times the space average of the emf per coil. Because the angular velocity is constant, the space average over this half of the periphery is equal to the time average of the emf induced in one coil over the time necessary to make one-half revolution.

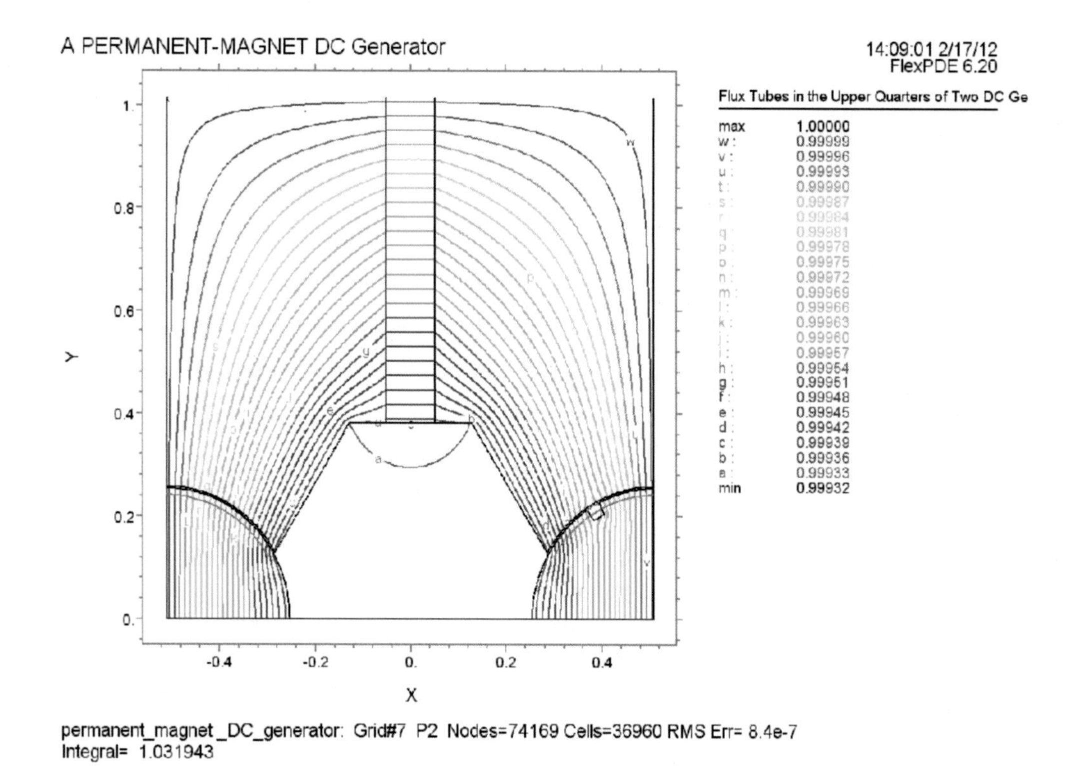

Figure 31: Flux tubes in the DC generators

Therefore the total generated voltage is $2N\phi_t/T$ where N is the number of coils on the armature. Because we are interested in only the average, the expression BLv is exact.

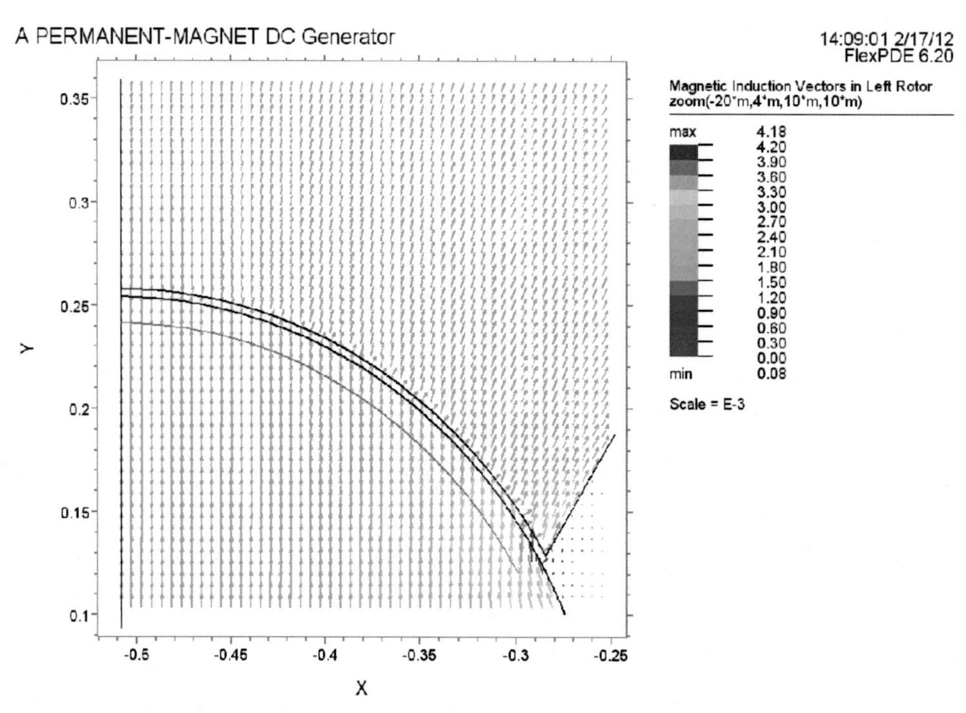

Figure 32: Magnetic induction vectors in the left rotor

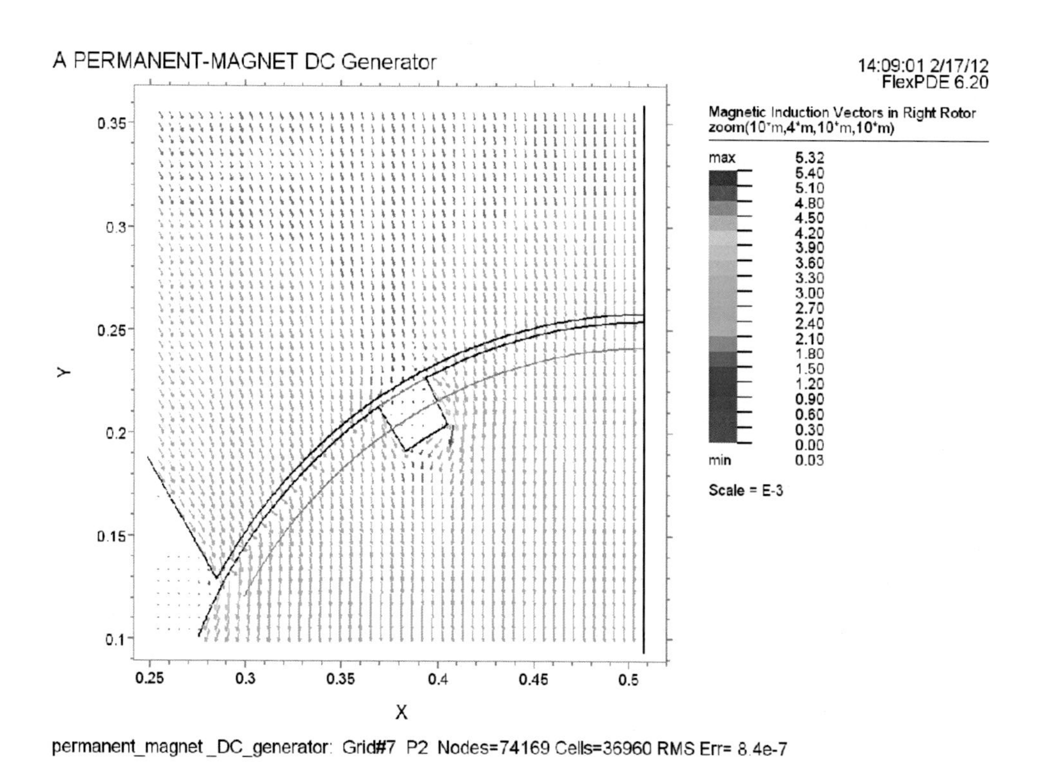

Figure 33: Magnetic induction vectors in the right rotor

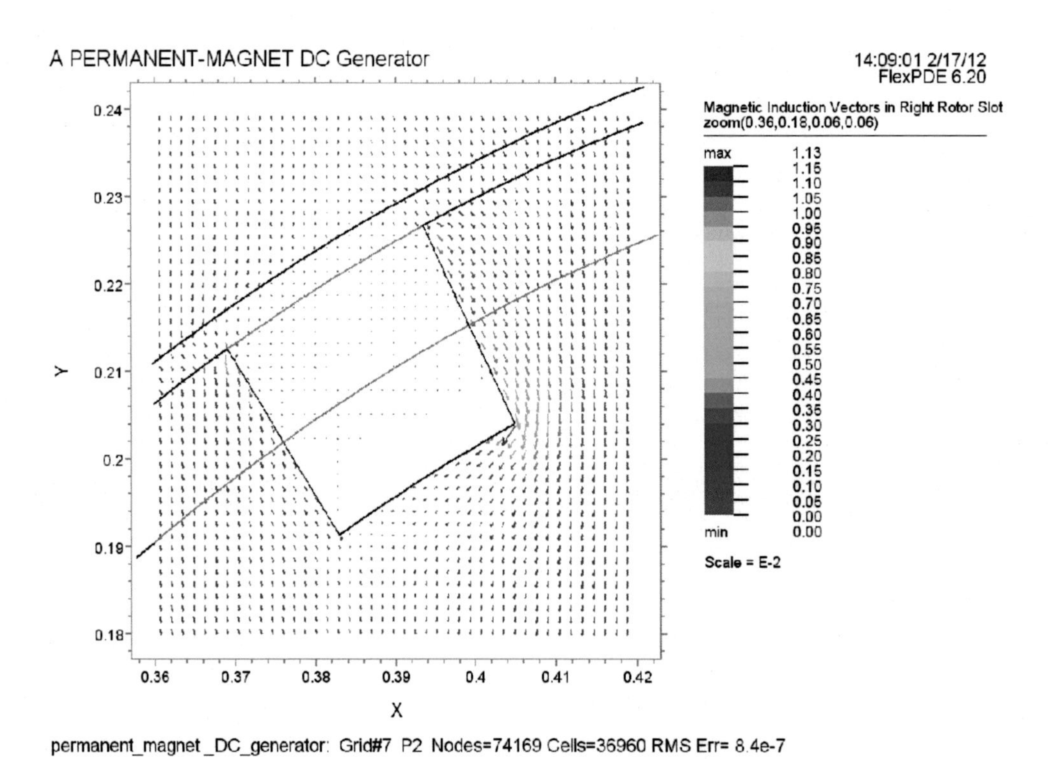

Figure 34: Magnetic induction vectors in the slot

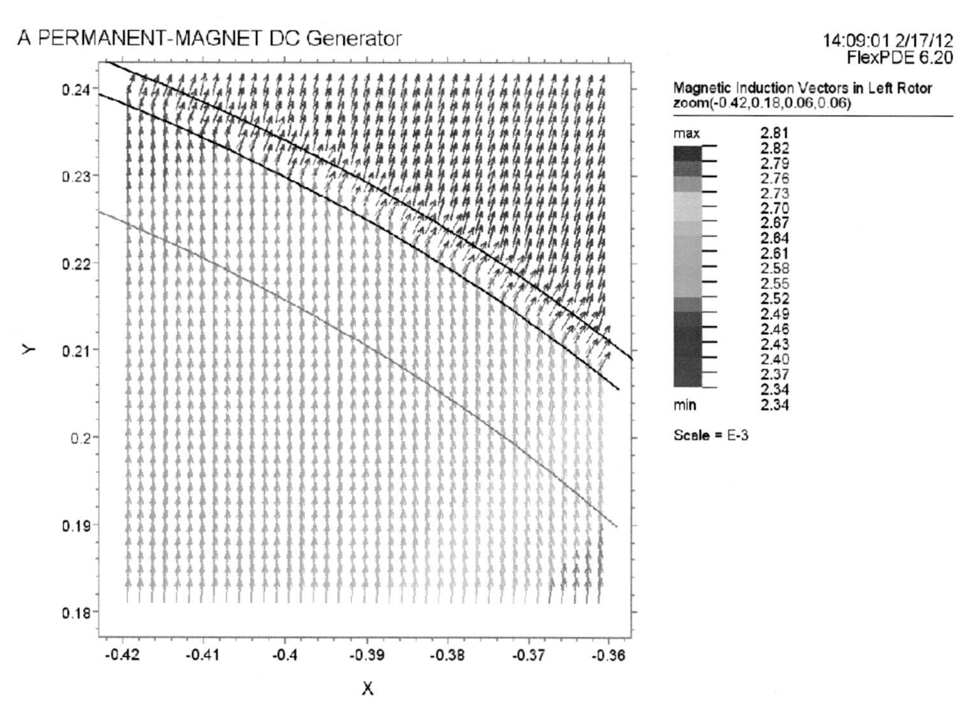

Figure 35: Magnetic induction vector (No slot)

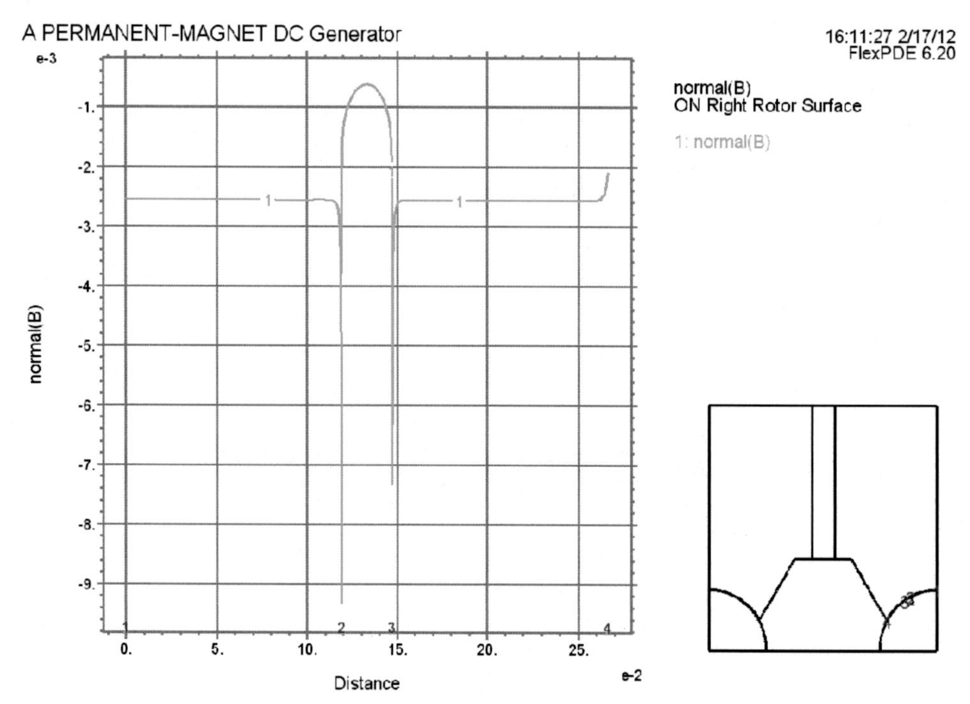

Figure 36: B_{normal} along the right rotor surface

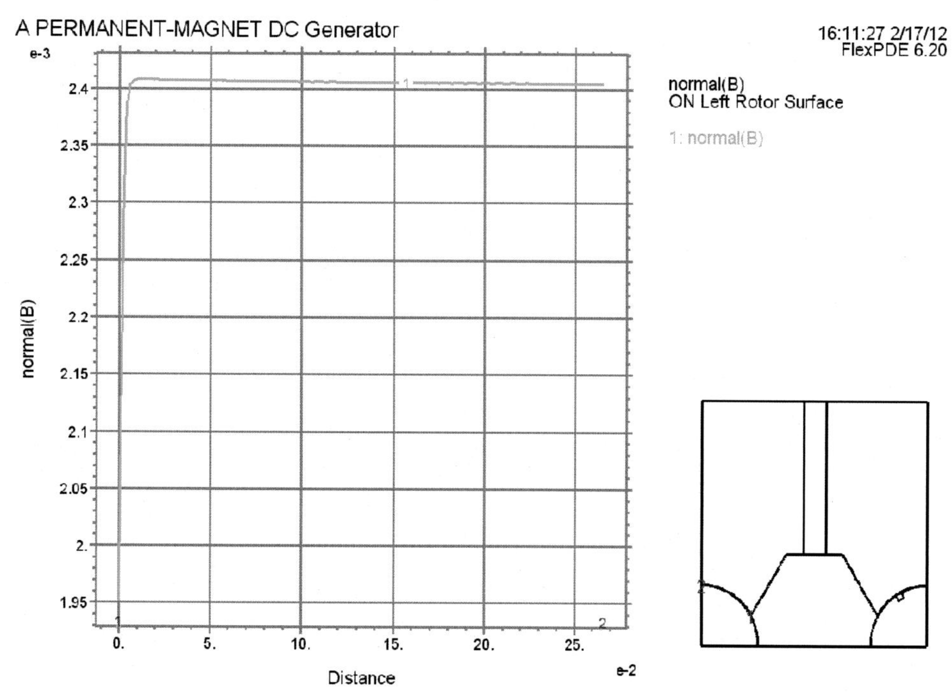

Figure 37: B_{normal} along the left rotor surface

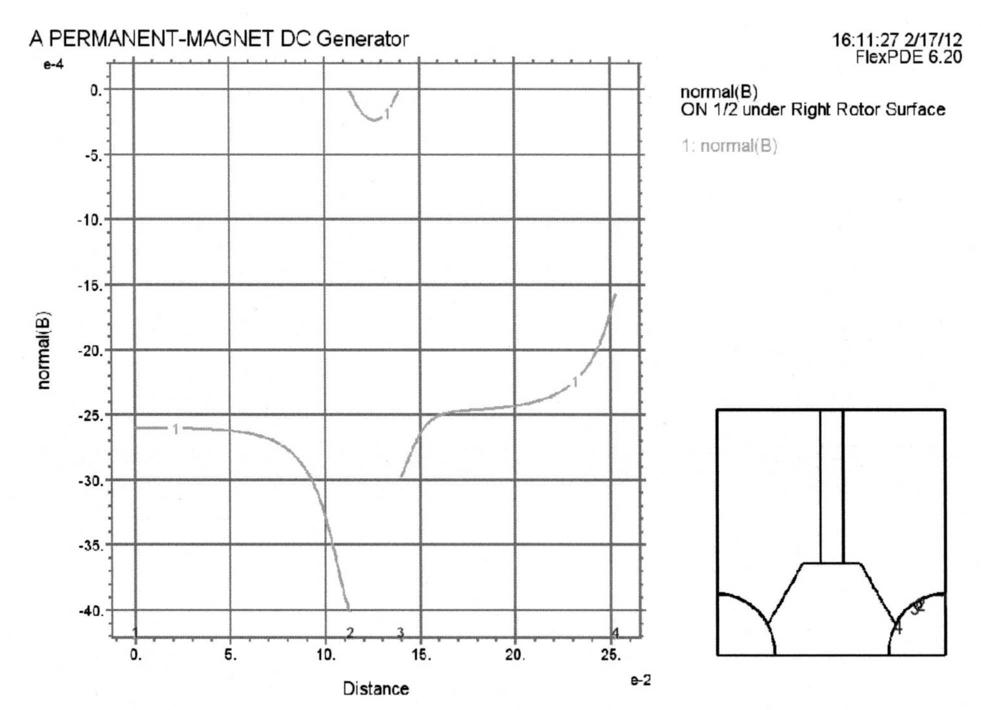

Figure 38: B_{normal} 1/2" under right rotor surface

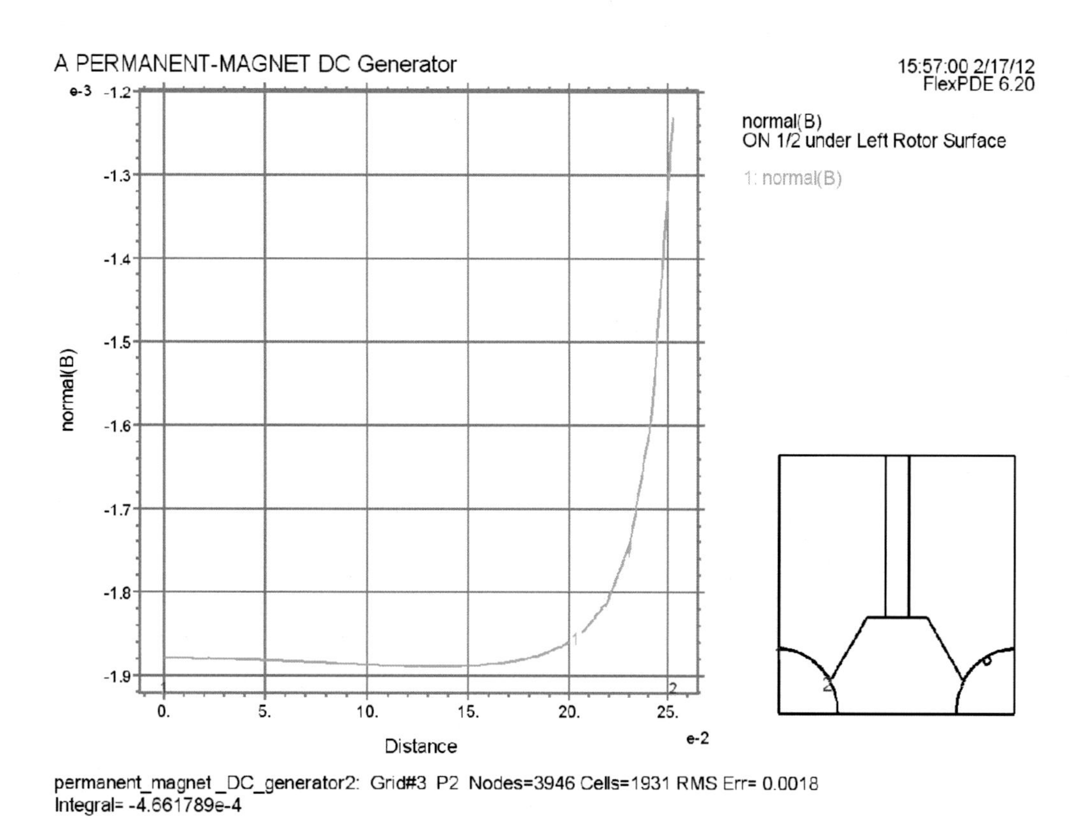

Figure 39: B_{normal} 1/2" under left rotor surface

6.13 Emf generated by plates rocking in a uniform magnetic field

In Fig.40 two copper plates with slightly curved edges are rocking back and forth between the solid and the dashed lines in a uniform magnetic induction field directed into the paper. Each plate is connected to one of the terminals of a voltmeter or galvanometer. The plates make contact at the point P initially making a complete circuit including the galvanometer. As the plates are rocked place of contact moves continuously from point P to point P'. If we imagine the circuit to be completed through the plates on the dotted line shown in the figure, the magnetic flux through this circuit changes by a large amount as the plates are rocked. Feynman, Leighton and Sands[8] wrongly claim this is an exception to Faraday's Law and show a complete lack of physical understanding of emf. Here they indicate that the circuit in the plates moves between the dotted and dash-dot lines and makes a $\mathbf{V} \times \mathbf{B}$ that can be minimized by making $\mathbf{V}$ very small. However, they say there may still be a large change in flux linking the circuit. The simplest way to solve for the emf is to place the observers on both plates. They will each say their plate is stationary and not moving in a magnetic field. However, they will notice the leads and galvanometer moving with respect to them. The observer on the left plate will see the right plate moving twice as fast as an observer on the galvanometer and he would see a small $2v \times \mathbf{B}$ in the connection made to the right plate and none in his connection. He would also see a small change in flux in the area between the dashed and dotted lines. This would lead to the very small potential Feynman et. al. observed. If the leads were stiff and the contacts were connected to brushes at the edge of each plate the leads would not move and there would be no emf at all generated. A slightly different configuration of these plates is shown in Fig.41. It has no leads that move as the plates are rocked. The line abcdefgh shows the base of the white plates at rest and that line is the most direct way for current to flow to the red brushes and around the loop through the voltmeter. There is a magnetic induction, $\mathbf{B}$ directed into the paper. The plates are moved approximately fifteen degrees and are depicted in that position by dashed lines and the part that would stick out were there two sets of plates is colored a light green. Initially the path from the voltmeter is given by abcdefgh. As the plates move from the initial position the path becomes abcije'klgh around the loop through the voltmeter. Along this path $-d\Phi/dt = 0$ The only point at which the right and left plates touch is at point e'. Faraday's Law can be applied if there are no breaks in the path. The $\mathbf{v} \times \mathbf{B} \cdot dl$ on the left plate cancels the same term on the right plate. Thus Eq.19 yields zero emf. An observer might be stationed on either the left or the right conducting plate or on the external circuit comprising the brushes and the voltmeter leads. As illustrated many times in this text, all observers must agree on the emf produced. The ones sitting on the plates would have to completely evaluate Eq.19 along a path and embraced area including the brushes and voltmeter leads. The observer on the external circuit, however, sees no time rate of change in the magnetic induction flux if he takes the correct path. His circuit goes counter clockwise

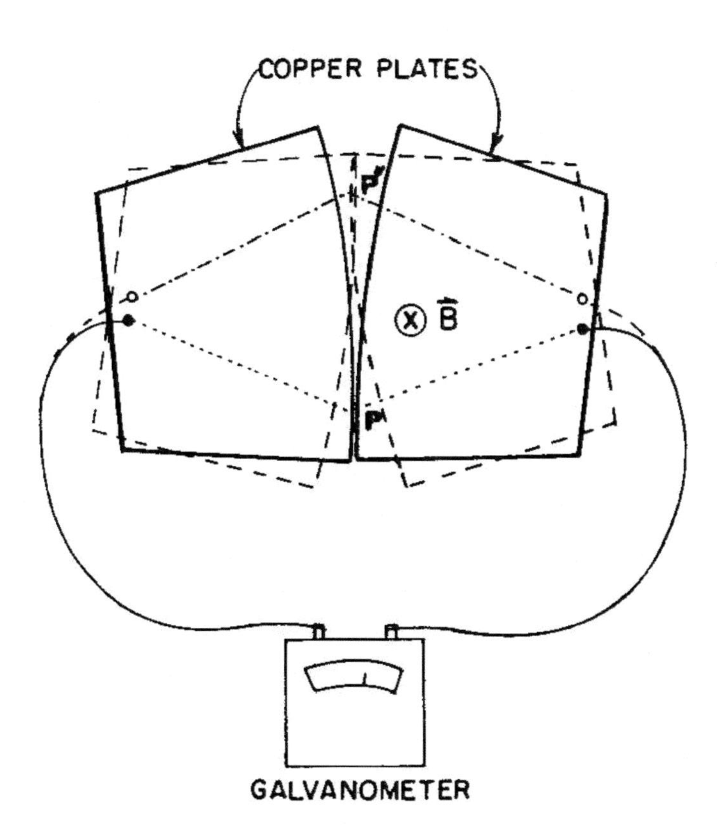

Figure 40: Feynman et al rocking plates

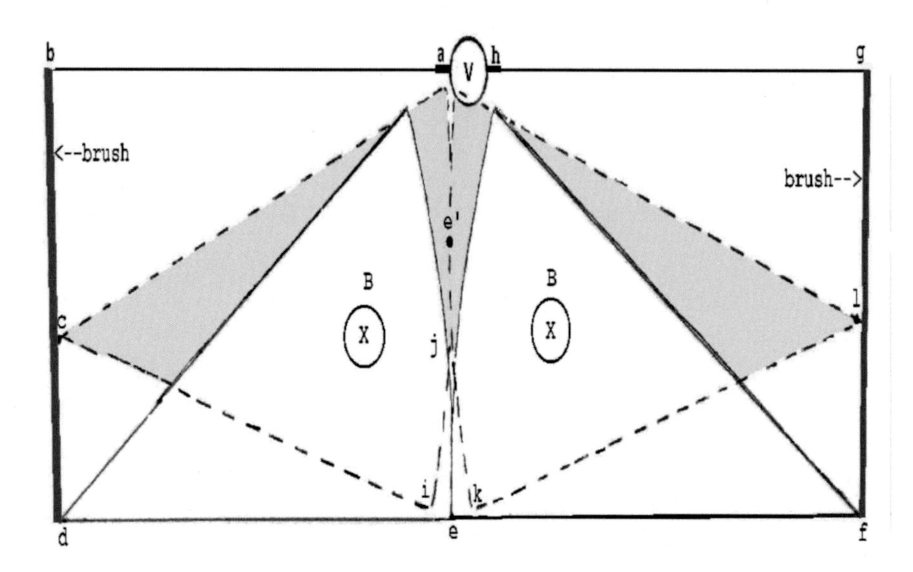

Figure 41: Two metal plates that rock

from the left terminal of the voltmeter to the left brush, passing through points abcdeflgh through the right brush to the right terminal of the voltmeter. There $e = -d\Phi/dt = 0$ regardless of the movement of the plates and independent of their magnetic relative permeability. This is the easy solution.

This text continues with several problems for the reader to consider. Try to pick the observer who finds the easiest solution.

7 Problems for reader solution

7.1 Emf in a bicycle clip drawn over an iron core containing flux

Fig.42 shows a spring metal clip drawn over an iron core excited by the circuit on the left side of the core. The clip is drawn so that it always contacts either itself or the iron core. Will there be any emf read by the voltmeter?

7.2 Emf in generated by a copper cylinder moving in a magnetic field

If a circular copper cylinder surrounds a current carrying conductor as shown in Fig.43 and has brushes on the inside and outside surfaces that are connected to a voltmeter, will there be an emf generated if the copper cylinder is moving

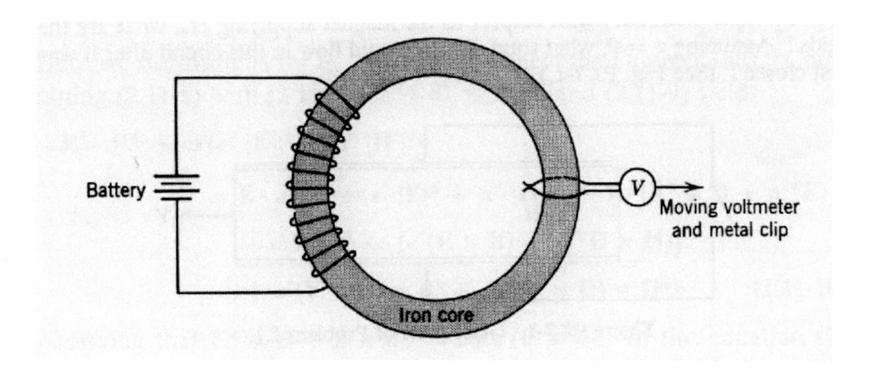

Figure 42: Configuration for Problem 1

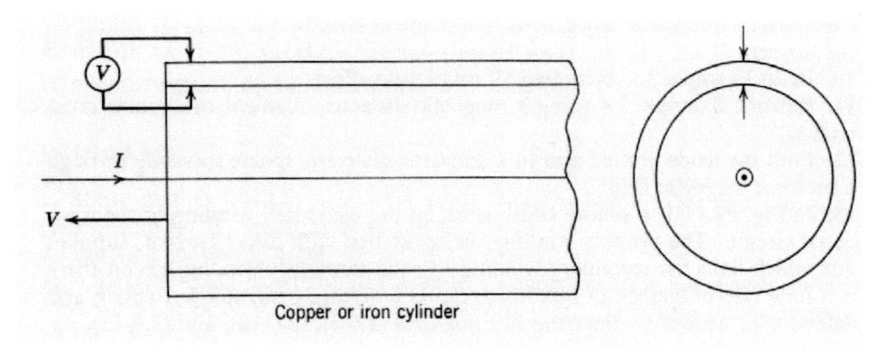

Figure 43: Configuration for problem 2

relative to the voltmeter with velocity, V? Would it make any difference if the cylinder were iron instead of copper?

7.3 Emf generated by a coil rotating around an excited magnetic core

If coil form is rotated by winding the wire on a reel as shown in Fig.44 will there be a voltmeter deflection in sketch (a)? Sketch (b)?

7.4 Emf generated by relative velocity between magnets

In Fig.45 are two disc-shaped magnets which have a hole drilled along their axes. They are magnetized so that the flat surfaces of the discs are magnetic poles. One disc is held fixed whilst the other rotates with an angular velocity Ω. Will there be an emf measured by the voltmeter?

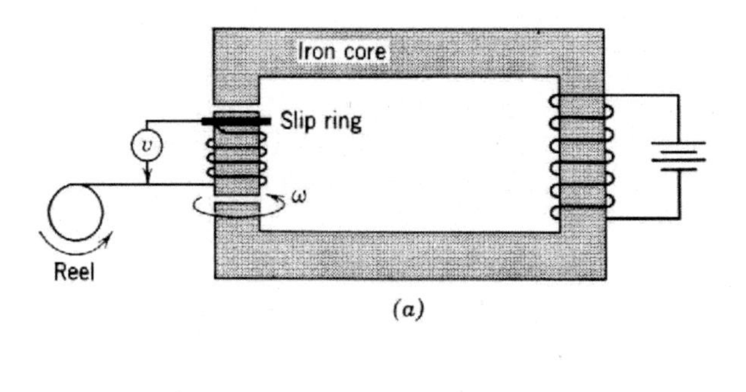

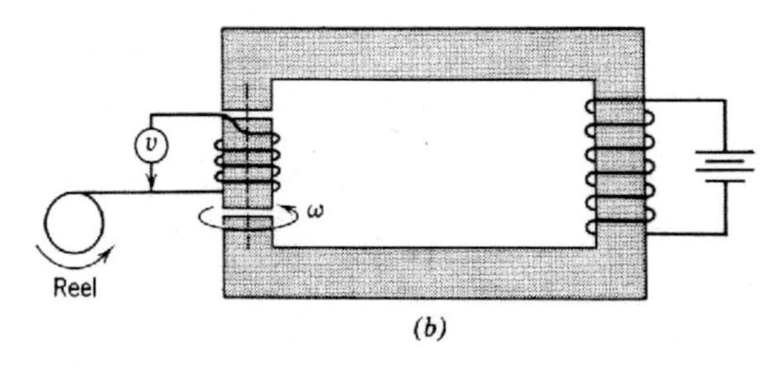

Figure 44: The configurations for problem 3

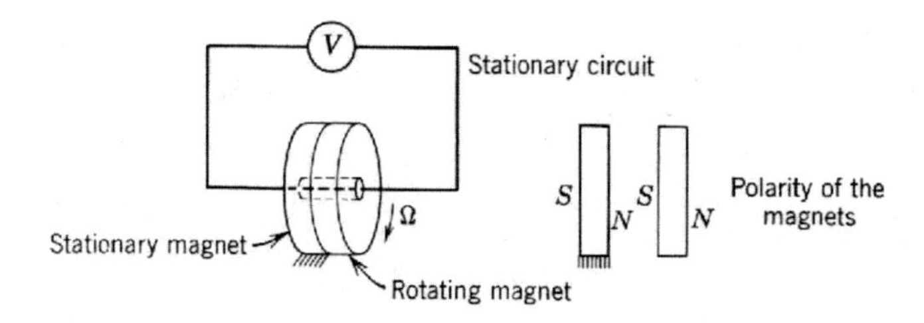

Figure 45: Configuration for problem 4

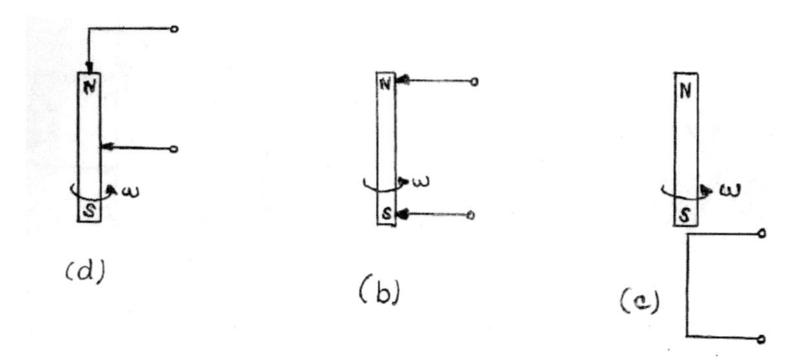

Figure 46: Configuration for problem 5

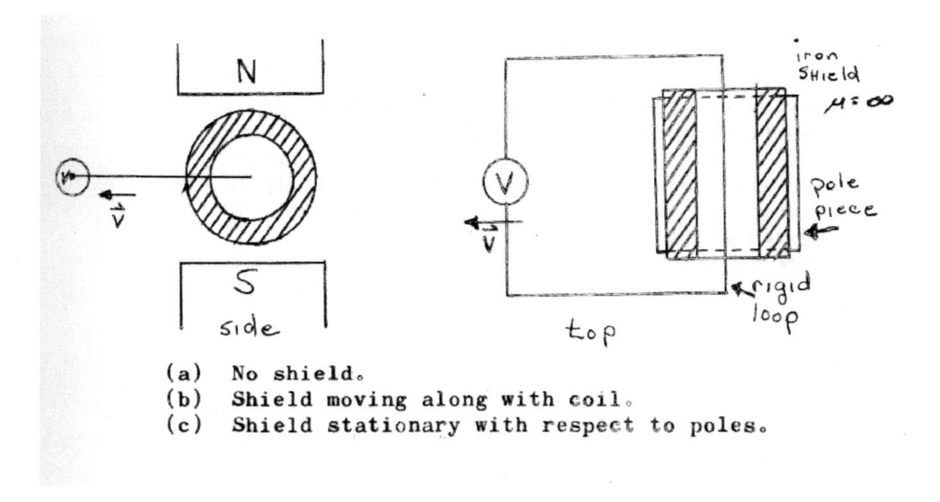

Figure 47: Configuration for problem 6

7.5 Emf generated by revolving magnets

Would there be a terminal voltage produced in any of the configurations shown in Fig.46 where the magnet is revolving on its own axis?

7.6 Emf produced by movement in a magnetic shield

Compare the emfs generated in the configurations shown in Fig.47.

7.7 Faraday generator with an anisotropic rotor

Is there an emf generated in the configuration shown in Fig.48? If so will it be pulsating direct current or alternating current?

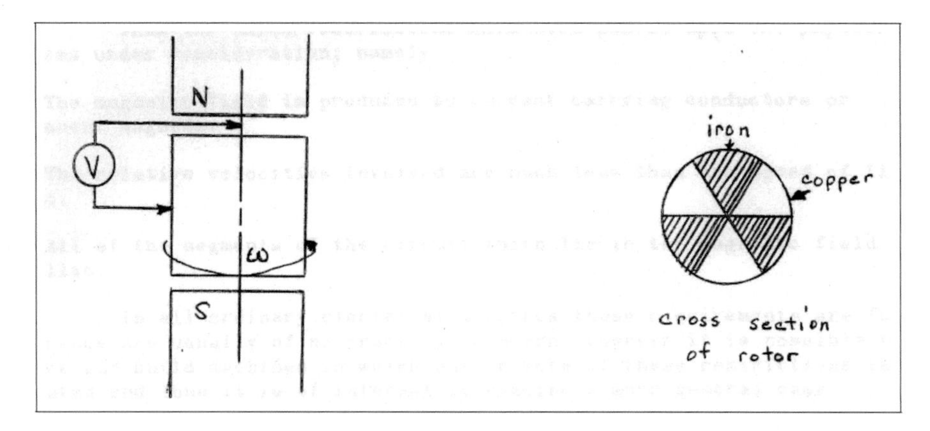

Figure 48: Configuration of problem 7

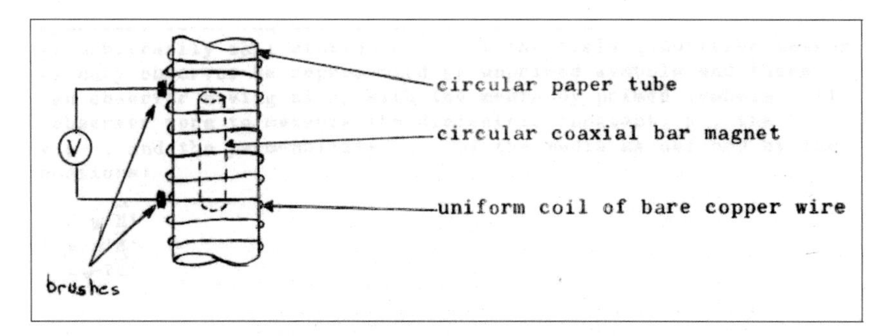

Figure 49: Configuration of problem 8

7.8 Bar magnet in a coil

In Fig.49 a circular bar magnet, magnetized along its axis, is centered in a bare copper coil wound on a paper tube. There is a voltmeter connected to the copper coil as depicted in said figure. Is there a voltmeter deflection

(a) if the bar magnet is moved along the axis,

(b) if the bar magnet is rotated about its axis,

(c) if the voltmeter and brush assembly is moved parallel to the axis of the coil such that the brushes slide from turn to turn without breaking the circuit,

(d) if the coil revolves in a helix about its axis such that the brushes make continuous contact along the wire and

(e) if the voltmeter and brush assembly is revolved in a helix such that the brushes make continuous contact along the wire?

Part VI
A MORE GENERAL FORMULATION

8 The Maxwell-Lorentz Transformation (MLT)

So far three restrictions have been placed upon the physical systems considered herein. They are

(a) The magnetic field is produced by current carrying conductors or permanent magnets

(b) The relative velocities involved are much less than the speed of light in vacuo

(c) All of the segments of the circuit which lie in the magnetic field are metallic.

In all ordinary electric machines these requirements are fulfilled and hence are usually of no practical concern; however it is possible to devise and build machines in which one or more of these restrictions is violated and thus it is of interest to examine a more general case.

As previously it is assumed that a given media (with any prescribed electric and magnetic properties) is moving with a velocity $\mathbf{V}$ relative to some observer whom is arbitrarily called stationary. Let the field quantities measured by this stationary observer be represented by unprimed symbols and those measured by an observer moving along with the media by primed symbols. If this latter observer were to measure the dielectric permittivity, ϵ'; the conductivity, σ'; and the permeability, μ', of the media as defined by $\mathbf{B}'=\mu'\mathbf{H}'$, $\mathbf{D}' = \epsilon'\mathbf{E}'$ and $\mathbf{J}'=\sigma'\mathbf{E}'$ he would *a posteriori* obtain the values ordinarily listed in the tables of properties of the particular media involved. That is true because the tables are compiled by such an observer. The stationary observer can define a similar set of values as $\mathbf{B} = \mu\mathbf{H}$, $\mathbf{D} = \epsilon\mathbf{E}$ and $\mathbf{J} = \sigma\mathbf{E}$. Here it is shown that these constants are in general not equal to the corresponding primed quantities. This means the material constants of a media are relative quantities and depend upon the relative speed between the media and the observer.

The key to the solution of this general case involving any arbitrary moving media lies in the following set of relations between the primed and the unprimed quantities. These equations are known as the Maxwell-Lorentz transformation equations and may be derived as a consequence of the Special Theory of Relativity. The derivation is done in numerous texts and appears in section 2.3 of Hughes and Young[9]. For the components perpendicular to the velocity there results

$$\mathbf{E}_{\perp}' = \beta(\mathbf{E} + \mathbf{V} \times \mathbf{B})_{\perp} \tag{63}$$

$$\mathbf{D}_{\perp}' = \beta(\mathbf{D} + \mathbf{V} \times \mathbf{H}/\mathbf{c}^2)_{\perp} \tag{64}$$

$$\mathbf{H}_{\perp}' = \beta(\mathbf{H} - \mathbf{V} \times \mathbf{D})_{\perp} \tag{65}$$

$$\mathbf{B}_{\perp}' = \beta(\mathbf{B} - \mathbf{V} \times \mathbf{E}/\mathbf{c}^2)_{\perp} \tag{66}$$

$$\mathbf{J}_{\perp}' = \beta(\mathbf{J}_{\perp} - \rho\mathbf{V}) \tag{67}$$

The subscript $\perp$ and $\shortparallel$ denote field components that are perpendicular and parallel to $\mathbf{V}$ and $\beta = (1 - V^2/c^2)^{-1/2}$. The quantity β is interesting because as $V^2/c^2 \longrightarrow 1$, $\beta \longrightarrow \infty$, thus making all the fields in the primed frame of reference very large. The components parallel to $\mathbf{V}$ are given by

$$\begin{vmatrix} \mathbf{E}_{\shortparallel}' \\ \mathbf{D}_{\shortparallel}' \\ \mathbf{H}_{\shortparallel}' \\ \mathbf{B}_{\shortparallel}' \\ \mathbf{J}_{\shortparallel}' \end{vmatrix} = \begin{vmatrix} \mathbf{E}_{\shortparallel} \\ \mathbf{D}_{\shortparallel} \\ \mathbf{H}_{\shortparallel} \\ \mathbf{B}_{\shortparallel} \\ \mathbf{J}_{\shortparallel} \end{vmatrix} \tag{68}$$

and

$$\rho = \beta(\rho - \mathbf{V} \cdot \mathbf{J}/c^2) \tag{69}$$

In these expressions the primed observer is moving with respect to the unprimed observer with a relative velocity $\mathbf{V}$ and is riding along with the medium. The inverse MLT () can be obtained by interchanging the primed and unprimed field quantities and substituting $-\mathbf{V}$ for $\mathbf{V}$. It may be important to realize that the MLT given here is strictly valid only when the frames are not accelerated. Otherwise the General Theory of must be used to obtain the correct transformations that become entangled with gravitational considerations involving the speed of light in vacuo. For accelerations obtainable in terrestrial devices such as motors, generators and certain weapon systems the above equations are good approximations to the truth. Moreover, in the derivation of these equations no constitutive medium is mentioned and therefore the MLT is valid for the most general kind of medium. Introducing the longitudinal stretcher operator given by

$$\alpha\mathbf{G} = \beta\mathbf{G}_{\shortparallel} + \mathbf{G}_{\perp} \tag{70}$$

allows the MLT to be written in a more compact form as follows

$$\begin{vmatrix} \mathbf{E}' \\ \mathbf{H}' \end{vmatrix} = \beta\left\{\begin{vmatrix} \mathbf{E}/\alpha \\ \mathbf{H}/\alpha \end{vmatrix} + \mathbf{V} \times \begin{vmatrix} \mathbf{B} \\ -\mathbf{D} \end{vmatrix}\right\} \tag{71}$$

and

$$\begin{vmatrix} \mathbf{D}' \\ \mathbf{B}' \end{vmatrix} = \beta\left\{\begin{vmatrix} \mathbf{D}/\alpha \\ \mathbf{B}/\alpha \end{vmatrix} + (\mathbf{V}/\mathbf{c}^2) \times \begin{vmatrix} \mathbf{H} \\ -\mathbf{E} \end{vmatrix}\right\} \tag{72}$$

and

$$\mathbf{J}' = \alpha(\mathbf{J} - \rho\mathbf{V}) \tag{73}$$

The charge density remains the same as given in Eq.69. In cases where the charge density, $\rho = 0$ in Eq.69 the $-\beta\mathbf{V} \cdot \mathbf{J}/c^2$ term cannot be neglected without causing $\nabla' \cdot \mathbf{D}' = \rho'$ to be invalid. The act of neglecting $\mathbf{V} \cdot \mathbf{J}/c^2$ is a fruitful source of paradoxes and should be done only after careful consideration.

9 The constitutive equations of media

So far it has not been important to say that S$'$or S is the laboratory frame of coordinates. Because the constitutive equations of the media are measured and are only true in the rest frame S$'$is chosen to be the rest frame and S the laboratory frame

9.1

In the electrodynamics of moving conducting media Ohm's law is important. A general statement valid in only the rest frame is $J_i' = \sigma_{ji} E_j'$, where the Cartesian tensor summation convention that $J_i' = \sum_{j=1}^{3} \sigma_{ji} E_j'$ for $i = 1,2,3$ is used. Substituting for J_i' and E_i' from Eq.71 and Eq.73 , respectively, into Ohm's law yields

$$J_i = (\beta/\alpha)\left\{\sigma_{ji}\left[E_j/\alpha + (\mathbf{V}\times\mathbf{B})_j\right]\right\} + \rho V_i \tag{74}$$

Here it is important to not confuse the order of the operators, since

$$(1/\alpha)[\sigma_{ji}E_j/\alpha] \neq \sigma_{ji}(E_j/\alpha^2)$$

Hence Ohm's law, which is relatively simple in the rest frame, becomes rather complicated in the laboratory frame. When $V^2 \ll c^2$Eq.74 becomes

$$J_i = \sigma_{ji}[E_j + (\mathbf{V}\times\mathbf{B})_j] + \rho V_i \tag{75}$$

which is valid for any anisotropic, nonlinear (for σ_{ji} might be a function of $\mathbf{E}$ and $\mathbf{B}$) medium traveling at a velocity small compared to the speed of light. Often in the study of there is very little charge density and the medium is such that $\sigma_{ji} = \delta_{ij}\sigma$. Then the medium is isotropic, e. g. water, mercury, NaK and various molten metals and Ohm's law in the laboratory frame becomes

$$\mathbf{J} = \sigma[\mathbf{E} + (\mathbf{V}\times\mathbf{B})] \tag{76}$$

Many authors take Eq.76 to be a fundamental principle, but it is really a result of the MLTs and a very special case of the general formulation given in Eq.74. One author[19], a former professor of mechanical engineering at Carnegie Institute of Technology did not realize the second term came from a cross product of two vectors and polled the literature to see which sign the term had in most papers. Often Ohm's law written in the laboratory frame is the only relationship that produces coupling between Maxwell's equation and the mechanical or fluidic momentum equations of the medium.

9.2 Dielectric permittivity and magnetic permeability

Consider the constitutive equations $D_i' = \epsilon_{ji}E_j'$ and $B_i' = \mu_{ji}H_j'$ which are valid in only the rest frame of the medium. The primed quantities in these relationships are replaced by using Eq.71 and Eq.72. There results

$$D_i/\alpha + (\mathbf{V} \times \mathbf{H}/c^2)_i = \beta\left[\epsilon_{ji}(E_j/\alpha) + \epsilon_{ji}(\mathbf{V} \times \mathbf{B})_j\right] \tag{77}$$

$$B_i/\alpha - (\mathbf{V} \times \mathbf{E}/c^2)_i = \beta\left[\mu_{ji}(H_j/\alpha) - \mu_{ji}(\mathbf{V} \times \mathbf{D})_j\right] \tag{78}$$

Eq.77 and Eq.78 together with Eq.71 and Eq.72 comprise six equations in eight unknowns: $E, D, B, H, E^{'}, D^{'}, B^{'}$, and $H^{'}$. Any one of these unknowns can be expressed in terms of any other two. This means that D cannot be expressed as a function of E only but must also depend upon another field quantity. For example, if a relationship between $D, E$and B is required H is obtained from Eq.78 and substituted into Eq.77 to yield

$$D_i = (1/\alpha - \alpha V^2/c^2)^{-1}\left\{\begin{array}{c} \epsilon_{ji}(E_j/\alpha) + \mathbf{V}/c^4 \times \alpha[\mu_{jk}^{-1}(\mathbf{V} \times \mathbf{E})_j] \\ +\epsilon_{ji}(\mathbf{V} \times \mathbf{B})_j - \mathbf{V}/c^2 \times \alpha[\mu_{jk}^{-1}(B_j/\alpha)] \end{array}\right\}_i \tag{79}$$

where $\mu_{jk}^{-1}$is the inverse of μ_{jk}. If $V^2/c^2 \ll 1$and if the medium is isotropic Eq.79 becomes

$$\begin{aligned} \mathbf{D} &= \epsilon^{'}\left\{\mathbf{E} + [\mathbf{1} - \mathbf{1}/(\mathbf{c}^2\mu^{'}\epsilon^{'})]\mathbf{V} \times \mathbf{B}\right\} \qquad (80) \\ &= \epsilon^{'}\left\{\mathbf{E} + [\mathbf{1} - \boldsymbol{\mu}_0\epsilon_0/(\mu^{'}\epsilon^{'})]\mathbf{V} \times \mathbf{B}\right\} \end{aligned}$$

It is noteworthy that all terms in the original equations containing c^{-2} were negligible in the final result. A similar expression for B in terms of H and D is derived by solving Eq.77 for E and substituting the result in Eq.78. The result is

$$B_i = (1/\alpha - \alpha V^2/c^2)^{-1}\left\{\begin{array}{c} \mu_{ji}(H_j/\alpha) + \mathbf{V}/c^4 \times \alpha[\epsilon_{jk}^{-1}(\mathbf{V} \times \mathbf{E})_j] \\ -\mu_{ji}(\mathbf{V} \times \mathbf{D})_j + \mathbf{V}/c^2 \times \alpha[\epsilon_{jk}^{-1}(D_j/\alpha)] \end{array}\right\}_i \tag{81}$$

If $V^2/c^2 \ll 1$ Eq.81 becomes

$$\begin{aligned} \mathbf{B} &= \mu^{'}\left\{\mathbf{H} + [\mathbf{1} - \mathbf{1}/(\mathbf{c}^2\mu^{'}\epsilon^{'})]\mathbf{V} \times \mathbf{D}\right\} \qquad (82) \\ &= \mu^{'}\left\{\mathbf{H} + [\mathbf{1} - \boldsymbol{\mu}_0\epsilon_0/(\mu^{'}\epsilon^{'})]\mathbf{V} \times \mathbf{D}\right\} \end{aligned}$$

In a classical experiment by Wilson and Wilson[5] Eq.80 was verified. In their experiment $\mathbf{E} = \mathbf{0}$ and since they used a material in which $\epsilon^{'}/\epsilon_0 = 6$ and $\mu^{'}/\mu_0 = 3$ this equation yields $\mathbf{D} = 17\mathbf{V} \times \mathbf{H}/c^2$; had they neglected the last term of Eq.81 the result would have been $\mathbf{D} = 18\mathbf{V} \times \mathbf{H}/c^2$. Eq.82 shows that even though a material in nonmagnetic to an observer stationary on it, the material shall appear to be magnet to an observer moving with respect to the material. In the limiting case where $\mu^{'} = \mu_0$ and $\epsilon^{'} = \epsilon_0$ then $\mathbf{B} = \mu_0\mathbf{H}$ and $\mathbf{D} = \epsilon_0\mathbf{E}$ from the last four equations. This agrees with the idea that all observers see the same magnetic and electrical characteristics of free space regardless of their relative velocities as dictated by the constant speed of light in vacuo.

To complete our discussion of the constitutive equations, we will determine the MLTs for the polarization and magnetization vectors. By the definition of $\mathbf{M}$, the magnetization vector, $\mathbf{B}' = \mu_0(\mathbf{H}' + \mathbf{M}')$ or $\mathbf{B} = \mu_0(\mathbf{H} + \mathbf{M})$ and B' and H' can be replaced in terms of the unprimed quantities by using Eq.71 and Eq.72. Then

$$\mu_0\mathbf{M} = (\beta/\alpha)(\mathbf{B} - \mu_0\mathbf{H}) + \mu_0\beta\mathbf{V}\times(\mathbf{D} - \epsilon_0\mathbf{E}) \tag{83}$$

which is easily simplified because the dielectric polarization, $\mathbf{P}$ is defined as $\mathbf{P} = \mathbf{D} - \epsilon_0\mathbf{E}$. The results are given by

$$\mathbf{M}' = \beta(\mathbf{M}/\alpha + \mathbf{V} \times \mathbf{P}) \tag{84}$$

and

$$\mathbf{P}' = \beta(\mathbf{P}/\alpha - \mathbf{V} \times \mathbf{M}/\mathbf{c}^2) \tag{85}$$

From the last two equations it is clear that a medium which is either a nonmagnetic dielectric or a nonpolarized magnetic material in its rest frame may appear to an observer in the laboratory frame as magnetic or polarized, respectively. This effect has been demonstrated experimentally by [6] and [7].

9.2.1 Conclusion

Conclusion 1 Note the very important fact that the so-called "motional field", $\mathbf{V} \times \mathbf{B}$ does not have the same polarizing action on the dielectric as the "static' field "$\mathbf{E}$". This is a very important concept and its appreciation will help to avoid all of the difficulties that one encounters if an attempt is made to separate the 'total" electric field intensity into its "static" and "motional" parts. Actually such a separation is not unique and in fact is quite meaningless and should be avoided. To any given observer there is only one $\mathbf{E}$ field and he cannot by any measurements determine whether this field results from free or bound charges or motion through a magnetic field or any combination thereof. All that can be said is there is an $\mathbf{E}$ field of a certain magnitude and direction.

9.3 Boundary conditions

Because Maxwell's equations are the same to all observers the boundary conditions that apply at the boundaries between media are also the same, regardless of relative velocities between observers. To illustrate the use of the MLT and the role of boundary conditions consider a lossy ferromagnetic dielectric slab traveling with velocity V_0 with respect to the xyz coordinate system through a uniform magnetic field normal to the slab as portrayed in Fig.50 and Fig.51. In the latter figure $z_0 \gg 2a$.

9.3.1 Approximate field solution

We wish to calculate the fields and charge distribution seen by the observer in the laboratory frame. Let subscripts 1 and 2 respectively denote free space and

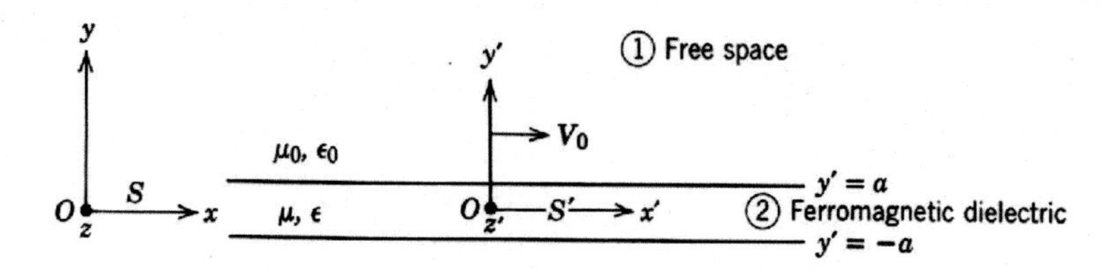

Figure 50: Moving lossy ferromagnetic dielectric slab

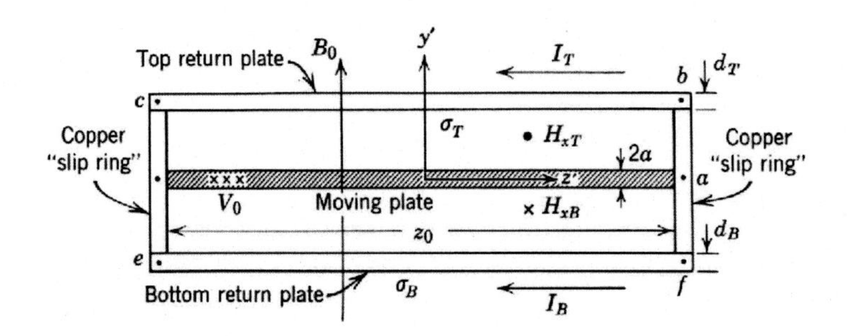

Figure 51: Moving slab and electrical return circuit

lossy ferromagnetic dielectric. Maxwell's equations written for the observer in frame S in both aforementioned figures are given from the divergence equations by

$$\partial B_y/\partial y = 0 \tag{86}$$

$$\partial D_y/\partial y = \rho \tag{87a}$$

because there are no time varying fields $\nabla \times \mathbf{E} = 0$ and because no space charge is assumed to exist in the rest frame

$$\partial E_z/\partial y = 0 \tag{88}$$

From the equation $\nabla \times \mathbf{H} = \mathbf{J}$ there results

$$-\partial H_{x1}/\partial y = 0 \tag{89}$$

$$-\partial H_{x2}/\partial y = J_{z2} \tag{90}$$

From Eq.73 $J_{z2} = J'_{z2} = \sigma E'_{z2}$and from Eq.71 E'_{z2} is known, so that Eq.90 becomes

$$-\partial H_{x2}/\partial y = \sigma(E_{z2} + V_0 B_{y2}) \tag{91}$$

So far it can be concluded that $B_{y1}, B_{y2}, H_{y1}, H_{y2}, D_{y1}, D_{y2}, E_{z1}, E_{z2}$ and $H_{x1}$are constants. By integrating Eq.91 H_{x2} becomes

$$H_{x2} = C_1 - \sigma(E_{z2} + V_0 B_0)y \tag{92}$$

for $y \geq 0$ and C_1 is a constant. The boundary condition $\mathbf{n} \times (\mathbf{H}'_1 - \mathbf{H}'_2) = \mathcal{F}$ known as the continuity of magnetic intensity at the boundary between media 1 and 2 is applied ($\mathcal{F}$ represents a current sheet at the boundary). The application of this boundary condition yields

$$H_{x1} = C_1 - \sigma(E_{z2} + V_0 B_0)a \tag{93}$$

where a is the half-width of the ferromagnetic slab. At this point in our analysis we know B_{y1}, H_{y1}, B_{y2} and we know H_{x1} in terms of $C_1$and E_{z2}. By the continuity of the tangential components of the electric intensity, $E_{z1} = E_{z2}$. So clearly $C_1$and E_{z2} must be found. Let $J'_{z2} = J_0$which depends upon the breadth of the ferromagnetic dielectric in the z direction and the method of electrical loading. Then

$$E_{z1} = E_{z2} = J_0/\sigma - V_0 B_0 \tag{94}$$

Fig.51 is a side view of the moving slab and its external connections. Here we assume the plate is so thin compared to the other dimensions that the assumption of a one-dimensional geometry is valid, and therefore end effects near the "slip rings" can be neglected. Let the top return circuit have a resistance per unit length of R_T and the bottom circuit R_B. Here $R_T = 1/(\sigma_T d_T x_0)$ and $R_B = 1/(\sigma_B d_B x_0)$, where x_0 is the length of the configuration in the x direction. The total current flowing is

$$I_0 = 2ax_0 J_0 \tag{95}$$

The current divides so that the current in the top and bottom return circuits is

$$I_T = 2ax_0 J_0[(\sigma_B d_B)^{-1}]/[(\sigma_T d_T)^{-1} + (\sigma_B d_B)^{-1}] \tag{96}$$

$$I_B = 2ax_0 J_0[(\sigma_T d_T)^{-1}]/[(\sigma_T d_T)^{-1} + (\sigma_B d_B)^{-1}] \tag{97a}$$

as dictated by Kirchhoff's voltage and current laws. Then

$$\mathbf{E}_T = \mathbf{E}_B = -\mathbf{z} \cdot 2aJ_0/(\sigma_T d_T + \sigma_B d_B) \tag{98a}$$

Because Kirchhoff's laws are compatible with Maxwell's equations, Eq.98a can be checked by the application of $\oint \mathbf{E} \cdot d\ell = 0$. Then

$$\int_{cb} \mathbf{E} \cdot d\mathbf{l} + \int_{fe} \mathbf{E} \cdot d\mathbf{l} = \mathbf{0}$$

This is clearly true since $\mathbf{E}_T = \mathbf{E}_B$ and the path elements are of opposite sign. Integration along path *dabcd* yields E_{z2} in terms of J_0. Performing the line integral yields

$$E_{z2} = -2aJ_0/(\sigma_T d_T + \sigma_B d_B) \tag{99}$$

If this value of E_{z2} is substituted into Eq.94, J_0 can be determined. This results in

$$J_0 = \sigma V_0 B_0 \{1 + 2a\sigma/(\sigma_T d_T + \sigma_B d_B)\}^{-1} \tag{100}$$

and

$$E_{z2} = -2a\sigma V_0 B_0/(\sigma_T d_T + \sigma_B d_B + 2a\sigma) \qquad (101a)$$

It is interesting to note that if either σ_T or σ_B is infinite, $J_0 = \sigma V_0 B_0$ and $E_{z2} = 0$. This is called the short-circuit case. When $\sigma_T = \sigma_B = 0$, $J_0 = 0$ and $E_{z2} = -V_0 B_0$ as would be expected from Eq.94. This is the open-circuit case. In the general case the currents are given by

$$I_T = 2ax_0\sigma_T d_T \sigma V_0 B_0/(\sigma_T d_T + \sigma_B d_B + 2a\sigma) \qquad (102)$$

and

$$I_B = 2ax_0\sigma_B d_B \sigma V_0 B_0/(\sigma_T d_T + \sigma_B d_B + 2a\sigma)$$

The terminal voltage is $V_T = I_T R_T = I_B R_B$. Then substituting either of the expressions for the current their results

$$V_T = 2az_0 V_0 B_0 \sigma/(\sigma_T d_T + \sigma_B d_B + 2a\sigma) \qquad (103)$$

To determine C_1 consider $\oint \mathbf{H} \cdot d\ell = \int_S \mathbf{J} \cdot \mathbf{dS}$ around a path that includes the ferromagnetic dielectric sheet and the bottom return circuit. Then

$$H_{x1}(y = a) = -2a\sigma V_0 B_0 \sigma_T d_T/(\sigma_T d_T + \sigma_B d_B + 2a\sigma) \qquad (104)$$

This result is easily checked. If $\sigma_B \to \infty$, the moving plate and the bottom return circuit comprise a solenoid outside of which the magnetic field is zero. So at $y = a$ it is to be expected that $H_{x1} = 0$. If $\sigma_T \to \infty$, then $H_{x1} = -2a\sigma V_0 B_0$, because the plane $y = a$ in now inside a solenoid. Thus Eq.104 is adequate in both cases. By using Eq.93 there results

$$C_1 = a\sigma V_0 B_0(\sigma_B d_B - \sigma_T d_T)/(\sigma_T d_T + \sigma_B d_B + 2a\sigma) \qquad (105)$$

The substitution of C_1 into Eq.92 yields

$$H_{x2} = [(\sigma_B d_B - \sigma_T d_T)a - ((\sigma_B d_B + \sigma_T d_T)y]\sigma V_0 B_0/(\sigma_T d_T + \sigma_B d_B + 2a\sigma) \qquad (106a)$$

To complete the solution for the observer in the laboratory frame H_{y1}, H_{y2}, D_{z1},and D_{z2} must be found. Clearly $H_{y1} = B_0/\mu_0$ and $D_{z1} = \epsilon_0 E_{z1}$. The electric displacement or induction in the ferromagnetic dielectric is obtained from Eq.80. The result is

$$D_{z2} = \beta^2 \epsilon V_0 B_0\{[1-\epsilon_0\mu_0/(\epsilon\mu)](\sigma_B d_B + \sigma_T d_T) - 2a\sigma\epsilon_0\mu_0/(\beta^2\epsilon\mu)\}/(\sigma_T d_T + \sigma_B d_B + 2a\sigma) \qquad (107)$$

Here it is noted that there exists values of load resistances that makes $D_{z2} = 0$ even though $E_{z2} \neq 0$. Setting $D_{z2} = 0$ yields

$$2a\sigma\beta^2 = (\kappa_e\kappa_m - 1)(\sigma_B d_B + \sigma_T d_T) \qquad (108)$$

whence is clear there are many ways to make $D_{z2} = 0$. It is easily shown from Eq.71 and Eq.72 that

$$H_{y2} = (\beta^2/\mu)[(1 - \epsilon\mu V_0^2) + (c^{-2} - \epsilon\mu)V_0 E_{z2}] \quad (109)$$

which upon the substitution of E_{z2} becomes

$$H_{y2} = (\beta^2/\mu)\{1 - \epsilon\mu V_0^2 - 2[(c^{-2} - \epsilon\mu)a\sigma V_0^2]/(\sigma_T d_T + \sigma_B d_B + 2a\sigma)\} \quad (110)$$

In this example we see that the return path for the current has an important effect on the field distributions. When $V_0^2 \ll c^2$, $H_{y2} \nrightarrow B_0/\mu$; but D_{z2} does not approach ϵE_{z2}. This may seem curious because the Maxwell Lorentz Transformations for $\mathbf{D}$ contains a V_0/c^2 but the MLT for $\mathbf{H}$ does not. However if the MLT is rewritten as $\mathbf{D}_\perp = \beta(\mathbf{D}' - \mathbf{V} \times \epsilon\mu\mathbf{H})_\perp = \epsilon\beta(\mathbf{E}' - \mathbf{V} \times \mathbf{B})_\perp$, then it is clear that $D_\perp$should not necessarily approach $\epsilon\mathbf{E}'_\perp$ when $V^2 \ll c^2$.

9.3.2 Terminal voltage without a field solution

Because $V_T = e - IR$ and $IR = I_0 z_0/(2\sigma a x_0)$ it follows that

$$V_T = z_0 V_0 B_0 - I_0 z_0/(2\sigma a x_0) \quad (111)$$

The current divides between the top and bottom plate such that $I_T = I_0 R_B/(R_T + R_B)$ and $V_T/R_T = I_T$. Therefore

$$I_0 = V_T(1/R_T + 1/R_B) \quad (112)$$

Accordingly, the substitution of I_0 from Eq.112 into Eq.111 yields

$$V_T = 2\sigma a x_0 V_0 B_0/[(z_0/x_0)(1/R_B + 1/R_T) + 2\sigma a] \quad (113)$$

where $1/R_B = \sigma_B d_B x_0/z_0$ and $1/R_T = \sigma_T d_T x_0/z_0$. Hence this is the same result obtained by the field solution given in Eq.103.

9.4 Terminal voltage related to the Poynting vector

The voltage rise from terminal 1 to 2 in Fig.52 is defined as $\mathrm{V}_{12} = -\int_1^2 \mathbf{E} \cdot d\ell$ or

$$V_{12} = \Phi_2 - \Phi_1 + \int_1^2 \partial\mathbf{A}/dt \cdot d\ell \quad (114)$$

Here $\mathbf{A}$ is the vector potential and $\int_1^2 \partial\mathbf{A}/dt \cdot d\ell$ is not independent of the path of integration. If the magnetic induction normal to the surface is zero for all

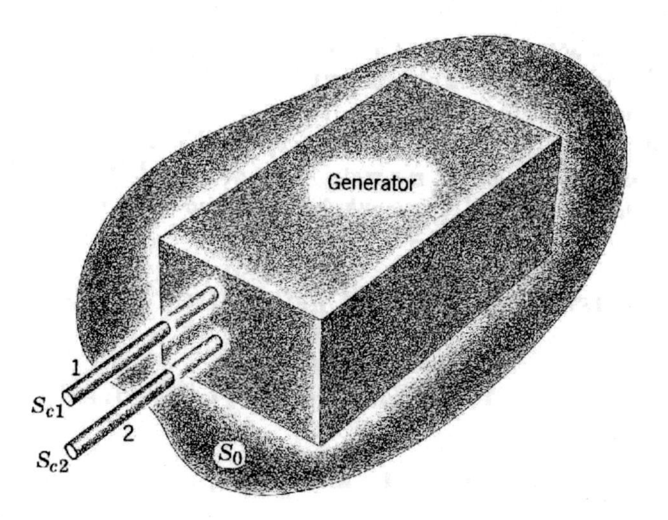

Figure 52: Closed surface surrounding generator

values of t, or if $\mathbf{E} \gg \partial\mathbf{A}/dt$, the voltage rise is $V_{12} = V_T = \Phi_2 - \Phi_1$. Where V_T is the terminal voltage which is independent of the position of the voltmeter leads as long as they are located on the plane S_0. Expressing the time rate of energy flow through the surface yields

$$\int_S (\mathbf{E} \times \mathbf{H}) \cdot d\mathbf{S} = -\int_S [(\nabla\Phi + \partial\mathbf{A}/dt) \times \mathbf{H}] \cdot d\mathbf{S} \tag{115}$$

If $|\mathbf{E}| \gg |\partial\mathbf{A}/dt|$, Eq.115 becomes

$$\int_S (\mathbf{E} \times \mathbf{H}) \cdot d\mathbf{S} = -\int_{vol} \nabla \cdot [\nabla \times (\Phi\mathbf{H})] dV + \int_S \Phi\nabla \times \mathbf{H} \cdot d\mathbf{S} \tag{116}$$

Because $\nabla \times (\Phi\mathbf{H}) \equiv 0$ Eq.116 becomes

$$\int_S (\mathbf{E} \times \mathbf{H}) \cdot d\mathbf{S} = \int_S \Phi(\mathbf{J} + \partial\mathbf{D}/dt) \cdot d\mathbf{S} \tag{117}$$

When the displacement current density, $\partial\mathbf{D}/dt$ is negligible compared to the conduction current, $\mathbf{J}$ Eq.117 reduces to

$$\int_S (\mathbf{E} \times \mathbf{H}) \cdot d\mathbf{S} = \int_{Sc1} \Phi\mathbf{J} \cdot d\mathbf{S} + \int_{Sc2} \Phi\mathbf{J} \cdot d\mathbf{S} \tag{118}$$

Assuming the potential is constant throughout each conductor, and that $\int_{Sc1} \Phi \mathbf{J}_1 \cdot d\mathbf{S} = -\int_{Sc2} \Phi \mathbf{J}_2 \cdot d\mathbf{S}$ Eq. 118 becomes

$$\int_S (\mathbf{E} \times \mathbf{H}) \cdot d\mathbf{S} = (\Phi_2 - \Phi_1) \int_{Sc2} \mathbf{J} \cdot d\mathbf{S} = V_T i \tag{119}$$

This expression relates the instantaneous time rate of energy flow into a pair of terminals to the instantaneous power leaving the generator. Eq.119 is valid to any observer agreeing that $|\partial \mathbf{A}/dt| \ll |\nabla \Phi|$ and that $|\partial \mathbf{D}/dt| \ll |\mathbf{J}|$. These restrictions may seem severe but are the same ones in effect when we speak of the terminal voltage of a transformer or alternating-current generator. In these cases the alternating fields in the vicinity are finite but small enough that $|\partial \mathbf{A}/dt| \ll |\nabla \Phi|$.

9.4.1 Terminal voltage of Fig.51 by use of the Poynting vector

Here is applied Eq.119 to recalculate the terminal voltage obtained by other means in previous examples. $\mathbf{E} \times \mathbf{H}$ is to be integrated over the surfaces in the xz plane, which are located just underneath and above the top and bottom return circuits, respectively. The parts of the integration over the yz and xy planes is zero because $\mathbf{E} \times \mathbf{H}$ is in the y direction. When applying Eq.119 care must be taken to integrate over a closed surface. By Ampere's circuital law

$$H_{xT} - H_{xB} = I_0/x_0 \tag{120}$$

and assuming no applied field in the x direction, $H_{xT} = -H_{xB}$. Then $H_{xT} = I_0/(2x_0)$, and H_{xT} and H_{xB} are directed as indicated in Fig.51. Underneath the top plate

$$\int_S (\mathbf{E} \times \mathbf{H}) \cdot d\mathbf{S} = \mathbf{y}(a\sigma V_0 B_0 I_0 z_0)/(\sigma_T d_T + \sigma_B d_B + 2a\sigma) \tag{121}$$

and above the bottom plate an identical result is obtained. Then by Eq.119

$$V_T = 2\sigma V_0 B_0 I_0 z_0/(\sigma_T d_T + \sigma_B d_B + 2a\sigma) \tag{122}$$

which agrees with the previous solutions to this problem.

9.5 Magnetohydrodynamic channel flow

When a conducting fluid flow in a channel exposed to a magnetic field currents and electric fields are induced as in the solid metals considered here. Magnetohydynamics is a very large and complicated field and one of the most elementary

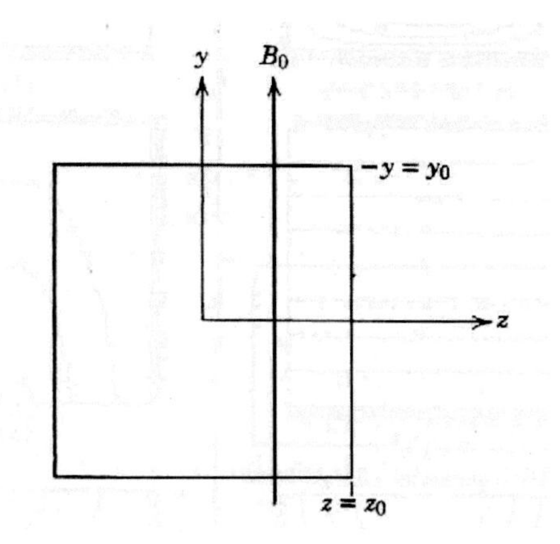

Figure 53: Coordinate system used in channel flow

cases are considered to show how electromotive force is calculated magnetohydrodynamic channels. Consider the case where the conducting fluid flows in the x direction and there is a constant and uniform magnetic field, B_0 applied in the y direction. This is depicted in Fig.53. Because $\mathbf{V} \times \mathbf{B}$ has a component in the z direction, it is expected that currents will flow in that direction causing a magnetic field to be induced in the x direction. Because the currents must have a return path there must generally be currents in the y direction. In fluid mechanics the fluid velocity in the x direction is often denoted as U in which case currents are caused by $\mathbf{U} \times \mathbf{B}$.When fluid enters a channel its velocity profile depends upon its source and as it flows down the channel it gradually changes into a steady profile that is the same along the length of the channel. That is called fully developed flow. The development also depends upon the applied magnetic field and is not considered here. We assume the flow is fully developed in the channel. We apply a constant pressure gradient to drive the fluid in the x direction. Our task is to find $\mathbf{J}, \mathbf{E}, \mathbf{H}, \mathbf{B}$ and U as functions of $x, y, B_0, \sigma, \mu_f, \mu_0$, x_0, y_0, z_0 and the appropriate boundary conditions. Thus the solution involves two of Maxwell's curl equations, the Navier-Stokes equation for the fluid dynamics and two constitutive equations. Because there are no time variations, the electric field is irrotational and $\nabla \times \mathbf{E} = 0$ becomes

$$\partial E_z / \partial y = \partial E_y / \partial z \tag{123}$$

Displacement current and space charge are neglected and Eq.76 is used in the $\nabla \times \mathbf{H} = \mathbf{J}$ equation. Then because the current density in the x direction is zero $\partial H_y / \partial z = 0$ and it is concluded that H_y is a constant given by $H_y = B_0 / \mu_0$ where B_0 is the constant magnetic induction applied to the channel. The remaining parts of the $\nabla \times \mathbf{H}$ equation yield

$$\partial H_x / \partial z = J_y = \sigma E_y \tag{124}$$

$$-\partial H_x / \partial y = \sigma(E_z + U B_0) \tag{125}$$

It is easy to combine the previous three equations to obtain

$$\partial^2 H_x/\partial y^2 + \partial^2 H_x/\partial z^2 + \sigma B_0 \partial U/\partial y = 0 \tag{126}$$

This is the convective part of the magnetic diffusion equation. Here the fluid is assumed to be incompressible, gravitational forces are neglected and there is no time variation in fluid velocity. Under these assumptions the Navier-Stokes equation becomes

$$0 = -\nabla P + \mu_f(\partial^2 \mathbf{V}/\partial y^2 + \partial^2 \mathbf{V}/\partial z^2) + \mathbf{J} \times \mathbf{B} \tag{127}$$

where P is pressure, $\mathbf{V}$ is the fluid velocity vector and the last term is the electromagnetic force retarding the fluid flow. Expanding the vector components of Eq.127 yields

$$0 = -\partial P/\partial x + \mu_f(\partial^2 U/\partial y^2 + \partial^2 U/\partial z^2) - J_z B_0 \tag{128}$$

$$0 = -\partial P/\partial y + J_z B_x \tag{129}$$

and

$$0 = -\partial P/\partial z + J_y B_x \tag{130}$$

From Eq.129 $\partial^2 P/\partial x \partial y = 0$ and from Eq.130 $\partial^2 P/\partial x \partial z = 0$, which is satisfied only when $\partial P/\partial x$ is a constant. In this problem the fluid is to flow in the x direction $\partial P/\partial x$ must be negative. Eq.129 and Eq.130 are useful for finding $\partial P/\partial y$ and $\partial P/\partial z$ after the current densities and magnetic inductions are known. Because $J_z = -\partial H_x/\partial y$ Eq.128 becomes

$$\partial^2 U/\partial y^2 + \partial^2 U/\partial z^2 + (B_0/\mu_f)\partial H_x/\partial y = (\partial P/\partial x)/\mu_f \tag{131}$$

Eq.126 and Eq.131 must be solved simultaneously subject to the appropriate boundary conditions.

9.5.1 Magnetohydrodynamic boundary conditions

There are four walls that can be various combinations of insulators, free surfaces or ideal conductors confining the fluid. Thus there are many arrangements yielding different boundary conditions. At conducting or insulating walls the fluid velocity is zero whereas for free surfaces the shear or normal derivative of velocity is zero. Mathematically these conditions are expressed as

$$U|_{boundary} = 0 \tag{132}$$

$$\partial U/\partial n|_{boundary} = 0 \tag{133}$$

When the walls are ideal conductors, the electric field intensity tangential to the walls must be zero or infinite current will flow in the walls. Hence at ideal conducting wall at rest in the laboratory frame of reference

$$\mathbf{n} \cdot \mathbf{J} = \partial H_x/\partial n|_{boundary} = 0 \tag{134}$$

At insulating walls the normal component of the current density must be zero and therefore

$$\mathbf{n} \cdot (\boldsymbol{\nabla} \times \mathbf{H}) = \partial H_x / \partial T|_{boundary} = 0 \tag{135}$$

where $\mathbf{n}$ is the unit vector normal to the boundary and these equations apply regardless of the velocity of the boundary. Eq.132 is known as the no slip condition that applies to a boundary at rest with respect to the fluid. If the boundary is in motions $U|_{boundary}$ = the boundary velocity.

9.5.2 Closed form solution of magnetohydrodynamic channels

Although there are analytical solutions to many different configurations of channels they are rather complicated and result in various infinite series of tabulated functions. Of most interest was the MHD generator configuration that was investigated by many authors and none could find an analytical solution. The materials problems encountered in building MHD generators were so severe that they are deemed unsuitable for domestic power generation. The trouble was the complete burning and melting of the electrodes after a few minutes of power generation. This did not concern USSR because MHD generators were to be used in remote areas where no electrical power was available. These generators were powered by hot electrically seeded gases, or plasmas that could be started by a radio signal very quickly to supply electrical power to missiles. The power was needed only a short time so it was of no importance that the MHD generator would be ruined by one launch. It seems that the MHD researchers in US took advantage of the lack of scientific knowledge of the US government to get many years of support to do fruitless research on MHD power generation. Under a National Science grant Stewart Way of Westinghouse Electric Corporation built and tested a small submarine powered by MHD channels of sea water passing through the ship. Sea water has an electrical conductivity of about 4 mho/m and the magnetic fields were less than 100 gauss. It moved so slowly that he could easily swim along side it. Perhaps it could be used as an auxiliary propulsion for a submarine to escape depth charges without making any engine noise. Many analytical solutions for magnetohydrodynamic channels are given by Hughes and Young[9]pp 195 - 258. The solutions to the partial differential equations given here are obtained by method used by Shercliff[20] in 1953. The results are summations of infinite series of tabulated functions satisfying the boundary conditions. They must be evaluated to produce numerical values and may involve careful computation to insure a specified accuracy. The solutions derived here are different than Shercliff's because they separate the variables differently but are equivalent. It is important to realize there are usually many ways to solve a given boundary value problem. The solution to MHD problems involving liquid metals is of great interest for cooling blankets expected to be used in thermonuclear reactors[21,22,23&24] when thermonuclear fusion becomes commercial.

For the geometry of Fig.53 it is clear that the velocity should be an even function of y whereas the induced magnetic intensity should be an odd function of y. In that case

$$U(z,y) = U_p + \sum_{n=0}^{\infty} f_n(z)\cos(\lambda_n y) \tag{136}$$

where U_p is a particular solution to Eq.126 and 131 and $\lambda_n = (2n+1)\pi/(2y_0)$. By integrating Eq.126 with respect to y, it is clear that the correct form for $H_x(z,y)$ is given by

$$H_x(z,y) = H_p + \sum_{n=0}^{\infty} g_n(z)\sin(\lambda_n y) \tag{137}$$

The assumed solutions must be adjusted to satisfy the appropriate boundary conditions. Because $U_0(\pm z_0, y) = 0$, the coefficients of the series must be given by $U_0(\pm z_0, y) = -\sum_{n=0}^{\infty} f_n(\pm z, y)\cos(\lambda_n y)$. Here $\lambda_n = (2n+1)\pi/(2z_0)$ and $U_p = U_{y_0\to\infty} = (z_0^2/2\mu_f)(\partial P/\partial x)[(z^2/z_0^2) - 1]$ which is the velocity profile in a channel having $y_0 \gg z_0$ and no applied magnetic induction, B_0. The particular solution, $H_p = 0$ because there are no induced fields when $B_0 = 0$. Then equations for $U(z,y)$ and $H_x(z,y)$ are homogeneous. When Eq.136 and Eq.137 are substituted into Eq. 126 and Eq.131 there results

$$d^2U_n(y)/dy^2 - \lambda_n^2 U_n(y) + (B_0/\mu_f)\, dH_n(y)/dy = 0 \tag{138}$$

and

$$d^2H_n(y)/dy^2 - \lambda_n^2 H_n(y) + \sigma B_0 dU_n(y)/dy = 0 \tag{139}$$

These equations are combined to yield and equation in $U_n(y)$ that is given by

$$d^4U_n(y)/dy^4 - \{2\lambda_n^2 + [M/(kz_0)]^2\}d^2U_n(y)/dy^2 + \lambda_n^4 U_n(y) = 0 \tag{140}$$

where $M = B_0 y_0(\sigma/\mu_f)^{1/2}$ and $k = y_0/z$. As the velocity profile should be the same for positive and negative values of y, the solution to this equation takes the form

$$U_n(y) = A_n ch(p_1 y) + B_n ch(p_2 y) \tag{141}$$

where

$$p_{1,2}^2 = \lambda_n^2 + [M/(kz_0)]^2/2 \pm [M/(kz_0)][\lambda_n^2 + [M/(2kz_0)]^2 \tag{142}$$

The magnetic intensity in the y direction is easily found from Eq.138 assuming $H_n(y=0) = 0$. Because all the generated current must flow inside the channel $H_n(\pm y_0) = 0$, automatically satisfying the condition that $(\partial H_x(y,z)/\partial z|_{\pm y_0} = 0$. Accordingly

$$A_n p_2(\lambda_n^2 - p_1^2)sh(p_1 y_0) + B_n p_1(\lambda_n^2 - p_2^2)sh(p_2 y_0) = 0 \tag{143}$$

The only unsatisfied boundary condition is $U(\pm y_0, z) = 0$. Expanding in a Fourier series yields

$$U_n(y_0) = [2(-1)^n/(\lambda_n^3 \mu_f z_0)]\partial P/\partial x \tag{144}$$

Applying the aforementioned boundary condition yields

$$A_n = [-p_1(\lambda_n^2 - p_2^2)/\Delta_n]U_n(y_0)sh(p_2y_0) \tag{145}$$

and

$$B_n = [-p_2(\lambda_n^2 - p_1^2)/\Delta_n]U_n(y_0)sh(p_1y_0) \tag{146}$$

where

$$\Delta_n = p_2(\lambda_n^2 - p_1^2)sh(p_1y_0)ch(p_2y_0) - p_1(\lambda_n^2 - p_2^2)sh(p_2y_0)ch(p_1y_0) \tag{147}$$

The solution for the velocity is

$$U(y,z) = (z_0^2/2\mu_f)\partial P/\partial x\{z^2/z_0^2 - 1 + (4/z_0^3)\sum_{n=0}^{\infty}(-1)^n\alpha_n(y)\cos(\lambda_n z)]/(\lambda_n^3\Delta_n) \tag{148}$$

where $\alpha_n(y) = -p_1(\lambda_n^2 - p_2^2)sh(p_2y_0)ch(p_1y) + p_2(\lambda_n^2 - p_1^2)sh(p_1y_0)ch(p_2y)$. It is easy to show that as $M \twoheadrightarrow 0$ Eq.148 reduces to the solution for viscous, incompressible channel flow. Having all boundary conditions satisfied the magnetic intensity is

$$H_x(y,z) = [2M/(kz_0^2)(\sigma/\mu_f)]\partial P/\partial x\sum_{n=0}^{\infty}(-1)^n\beta_n(y)\cos(\lambda_n z)]/(\lambda_n\Delta_n) \tag{149}$$

where $\beta_n(y) = sh(p_2y_0)sh(p_1y) - sh(p_1y_0)sh(p_2y)$. In magnetohydrodynamic problems the closed form solutions tend to be intricate and tedious to evaluate numerically. Care must be taken to make sure enough terms are used in the series to give accurate results and the Reynolds number should not be more than about 2000 or turbulence will result. The solutions given here are good for laminar flow only. The closed form solutions should disclose physical trends from inspection. Here the most obvious trend is that both velocity and magnetic intensity are directly proportional to the pressure gradient. Numerical calculations of the velocity profile and the induced magnetic intensity are given in detail by Hughes and Young[9] on pages 209 to 215. In mid 20th century these calculations were very long and tedious and the curves had to be hand plotted by draftsmen. Presently it is easier to obtain results by boundary or finite element calculations. Because the geometry is closed or does not extend beyond the channel walls finite element calculations are used.

9.5.3 Finite element calculation of magnetohydrodynamic channel with insulating walls.

There exist numerous commercial flow meters that measure the flow of fluids by MHD flow meters. One of the earliest papers on that subject was written by Thürlemann[25] in 1941. The flow velocities of fluids ranging from blood, sewage and slurries of stone and sand are determined by measuring the induced emf. Here the emf is calculated from the finite element solution of the Navier-Stokes and Maxwell equations.

Eq.126 and 131 are solved numerically subject to the no slip condition $U_x = 0$ and the condition that the tangential derivative $\partial H/\partial T = 0$, at the walls. Although the software has no provision for such a boundary condition it can be realized by using $H = 0$ at the walls. This is the case because all magnetic fields produced in the channel result from circulating currents and therefore must be internal to the channel. The fluid used is liquid lithium at 200°C. and the channel walls have zero electrical conductivity. The equations used are those derived here. The script for solving this problem is given in Appendix H. The geometry and boundary conditions are included in Region 1 under boundaries. Several plots of velocity, magnetic intensity, current density and electric field are requested. For a given amount of fluid flow it is desired to find the emf induced as a function of the Hartmann number . Here the pressure gradient, Pg is calculated to produce the same flow for all Hartmann numbers. Pg must get larger as the Hartmann numbers increases because it is more difficult to push the fluid against the increasingly large induced $\mathbf{J} \times \mathbf{B}$. Several of the plots obtained using this script are presented below.

Fig.54 is the case of ordinary viscous fluid channel flow ($M = 0$) because there is no magnetic induction in the y direction. The next figure has a magnetic induction applied producing a Hartmann number of unity. There is not much difference between the velocity profiles of M = 0 and M = 1. Fig.126 shows the velocity profile for M = 1.291549665. The next five figures exhibit the velocity profiles for M = 4.641588834, 10 12.91549665, 46.41588837, 100. The Hartmann numbers are chosen to be logarithmically spaced. The average fluid velocity is kept constant for each Hartmann number by an appropriate increase in the pressure gradient driving the flow. This is done by the use of a global variable that is solved in addition to the fluid velocity and magnetic intensity. Comparison of the velocity profiles shows the flow becoming more slug-like as the Hartmann number increases.Comparison of the contours of the magnetic intensity as a function of Hartmann number indicates that the current flow lines straighten and the concentrate near the extrema of y as the Hartmann number increases. Physically the current flows in the channel center as though the vertical walls are conductors. As the current nears those walls it must gradually change direction and go back along the horizontal walls. This is illustrated in Fig.69. A magnification of the lower left corner is exhibited in Fig.70. The script for this configuration yields figures like the aforementioned for each value of M. If more stages are defined so that M changes more gradually, the software generates movies.

Fig.71 shows the concentration of return current near the y = 0.01 plane when the Hartmann number is one hundred. The emf produced from the center of the channel at the plane y = 0 is shown in Fig.72. It grows quickly as Hartmann number increases through small values and begins to saturate around $M = 12$. The current in the z direction crossed into the applied magnetic induction, B_0 in the y direction produces an electromagnetic force that both retards and accelerates the liquid lithium in different parts of the channel. This is clear from Fig.73 and Fig.74. This force holds back the fluid in the core of the channel whilst accelerating it at the top and bottom of the channel. The

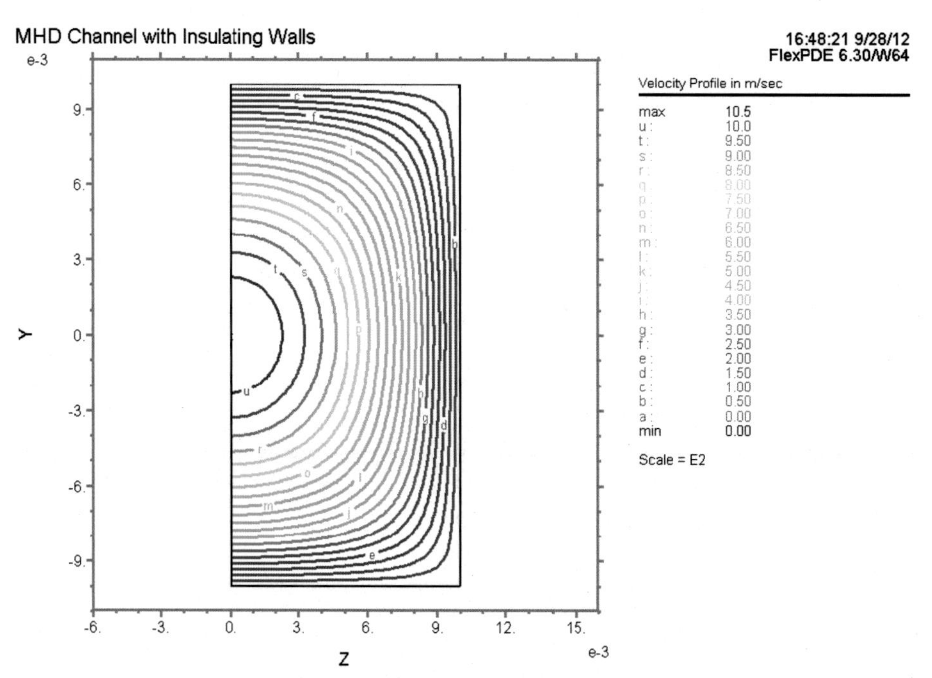

Figure 54: The velocity profile for M = 0

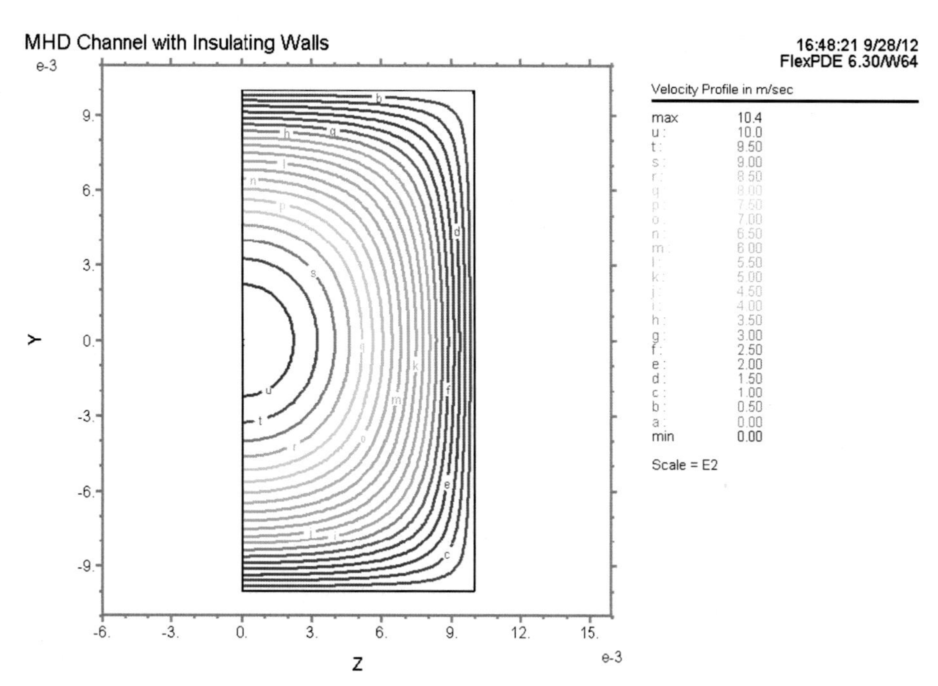

Figure 55: Velocity profile for M = 1

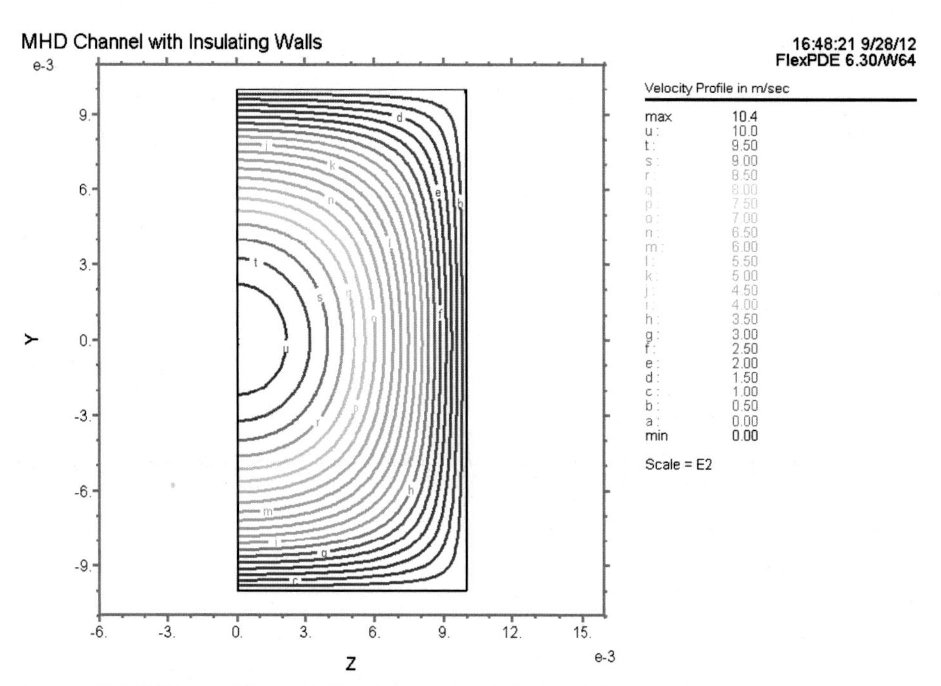

Figure 56: Velocity profile for M = 1.29155

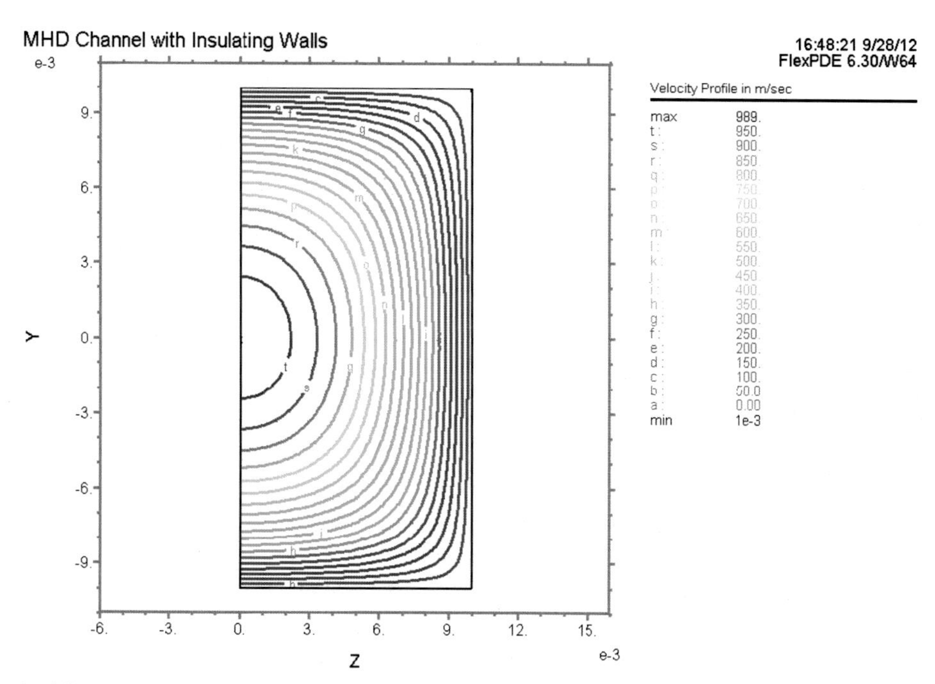

Figure 57: Velocity profile for M = 4.641589

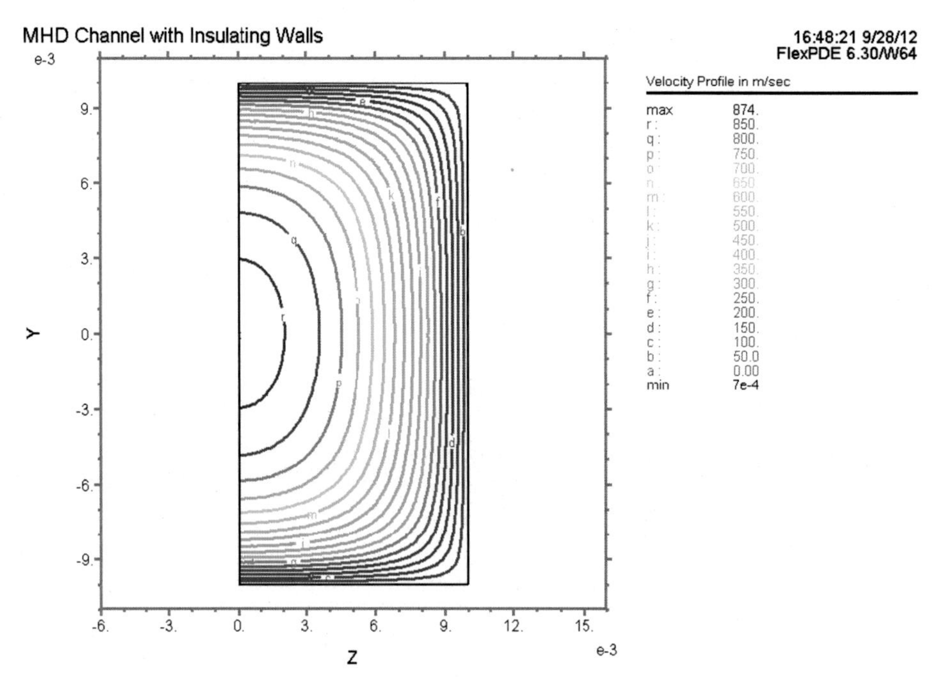

Figure 58: Velocity profile for M = 10

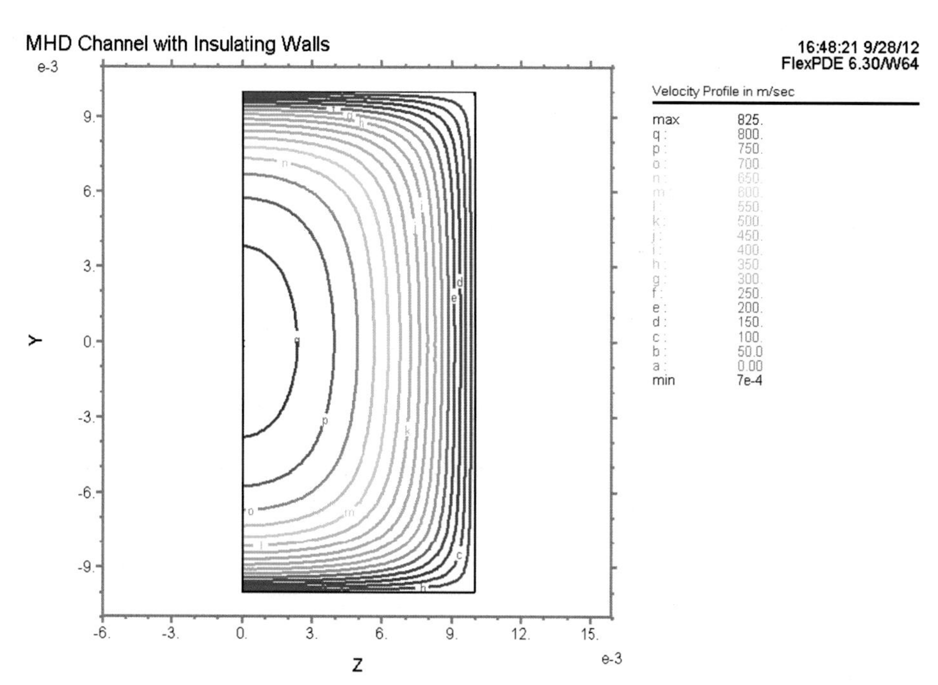

Figure 59: Velocity profile for M = 12.9155

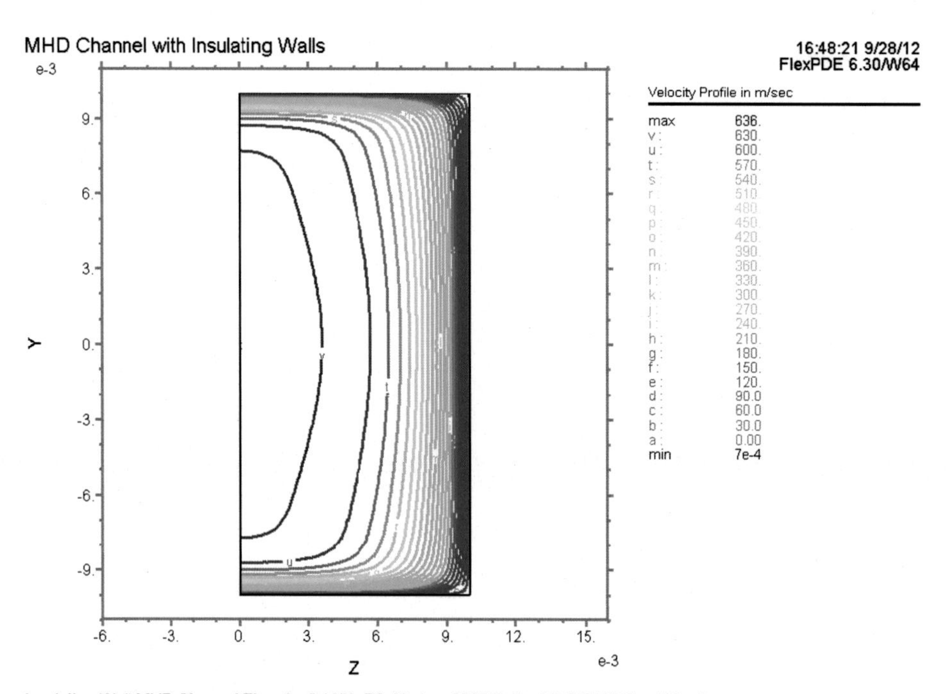

Figure 60: Velocity profile for M = 46.41589

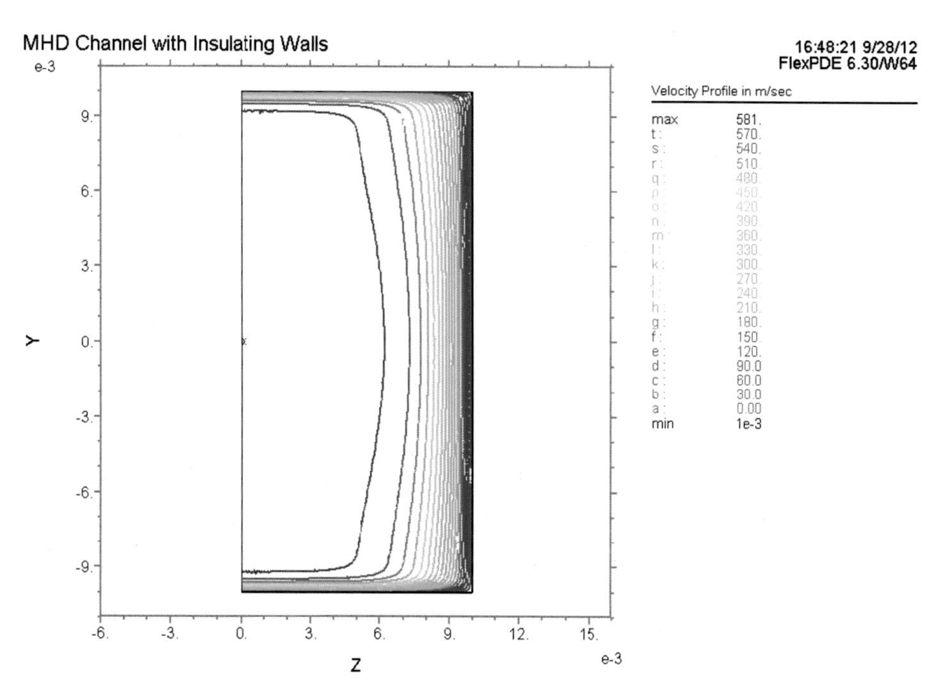

Figure 61: Velocity profile for M = 100

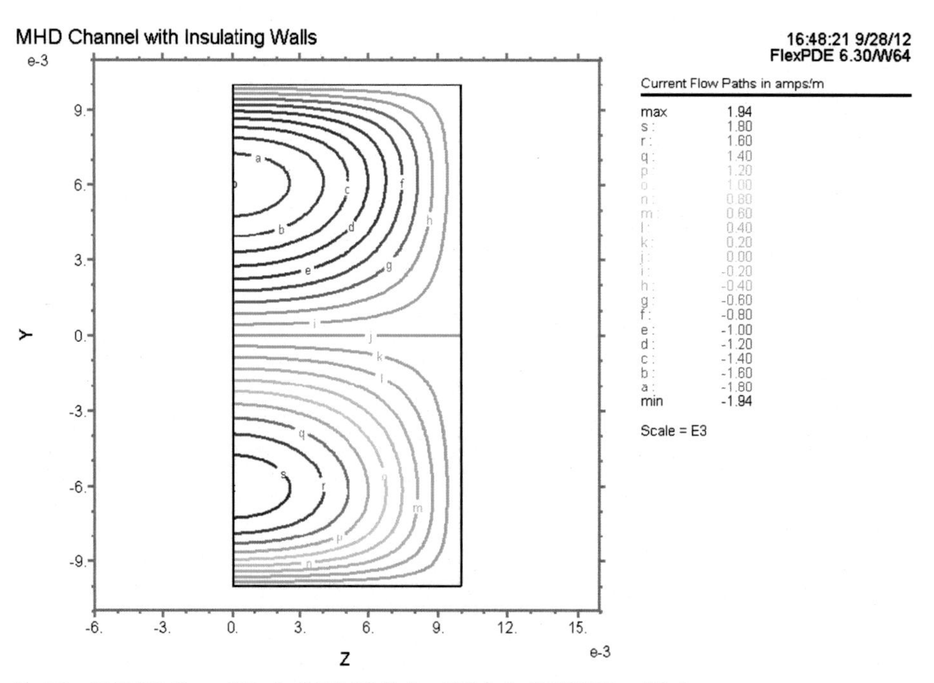

Figure 62: Magnetic intensity showing current paths

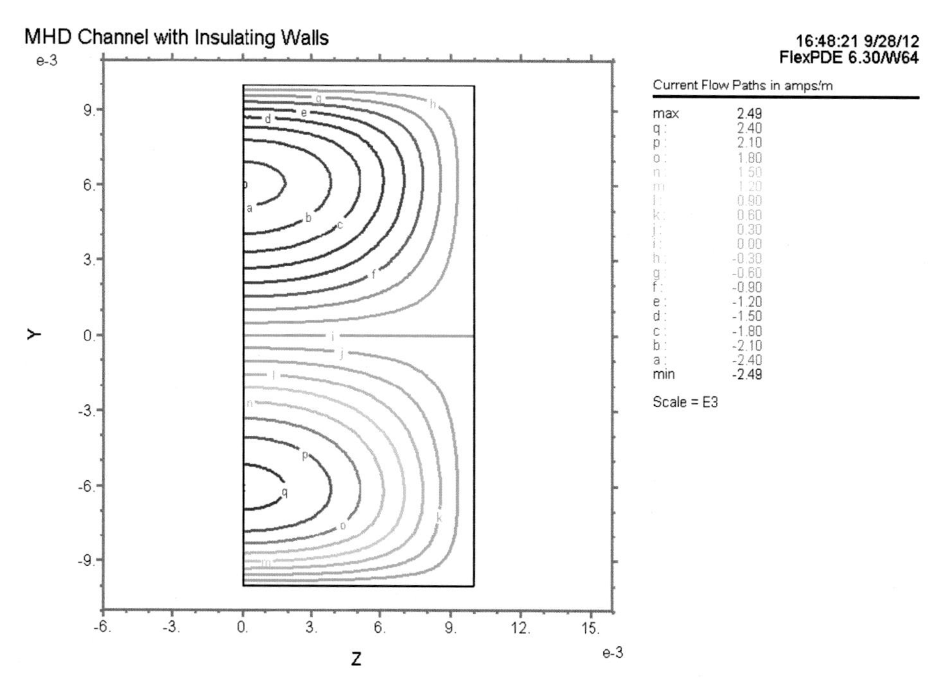

Insulating Wall MHD Channel Flow 4: Grid#1 P2 Nodes=1233 Cells=580 RMS Err= 8.8e-5
Stage 3 M= 1.291550 B0= 1.056929e-3 Pg= 21261.39 Integral= -4.771102e-6

Figure 63: Magnetic intensity for M = 1.291550

95

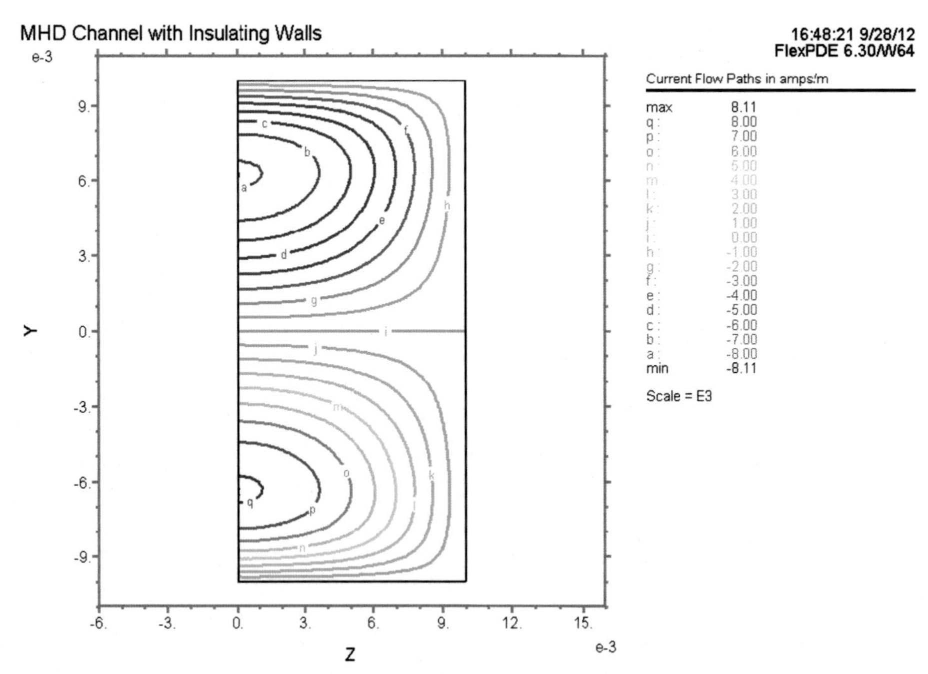

Figure 64: Magnetic intensity for M = 4.641589

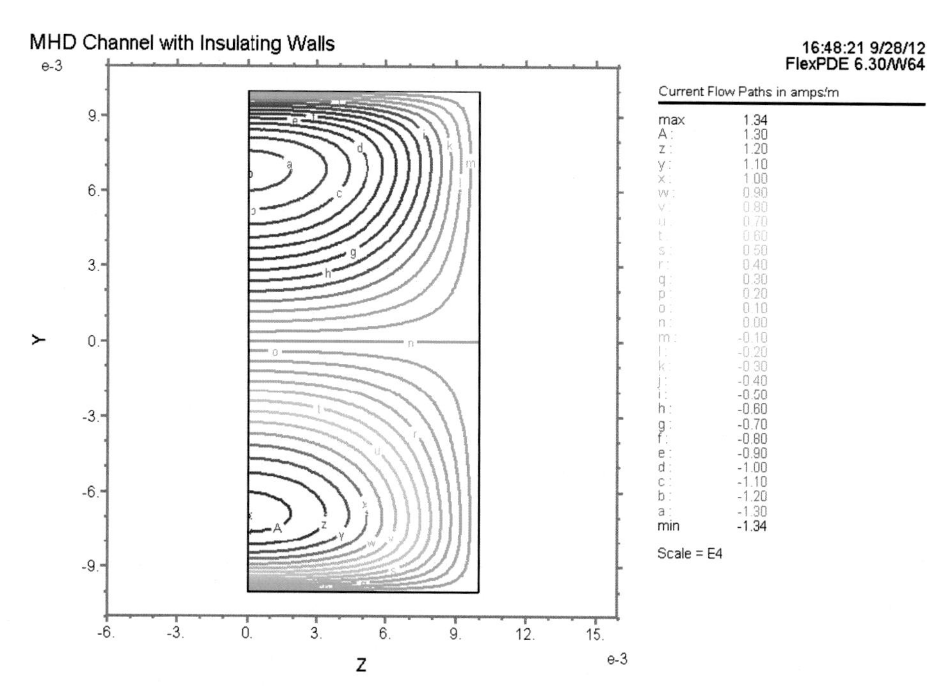

Figure 65: Magnetic intensity for M = 10

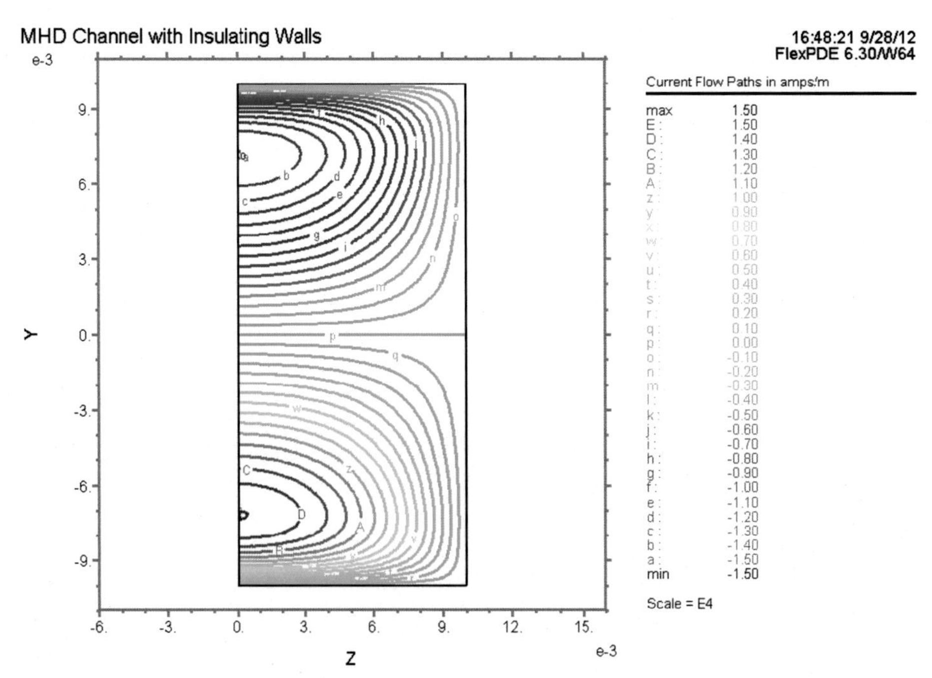

Figure 66: Magnetic intensity for M = 12.9155

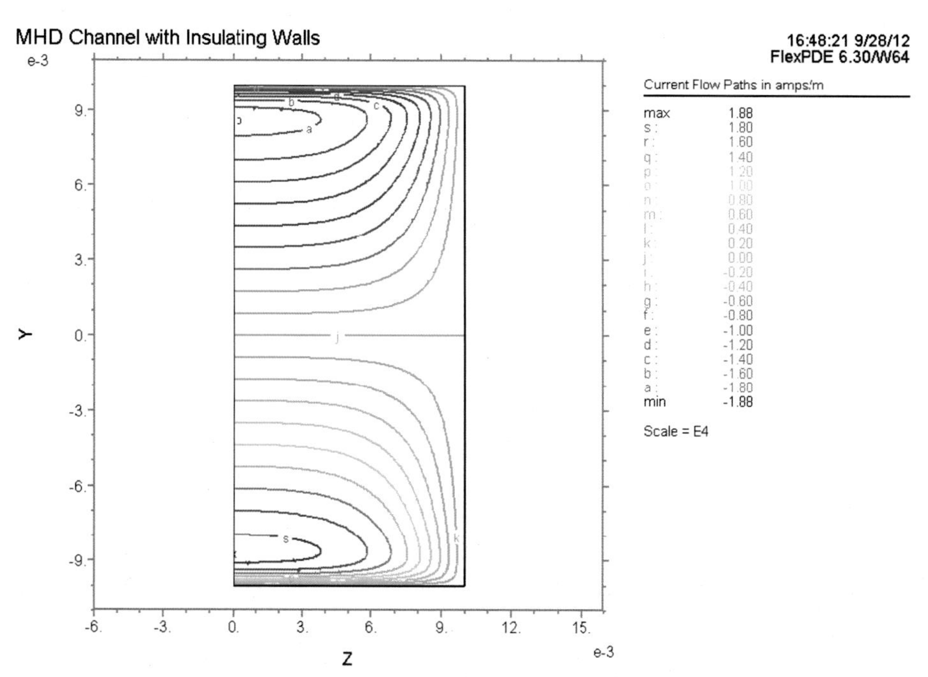

Figure 67: Magnetic intensity for M = 46.41589

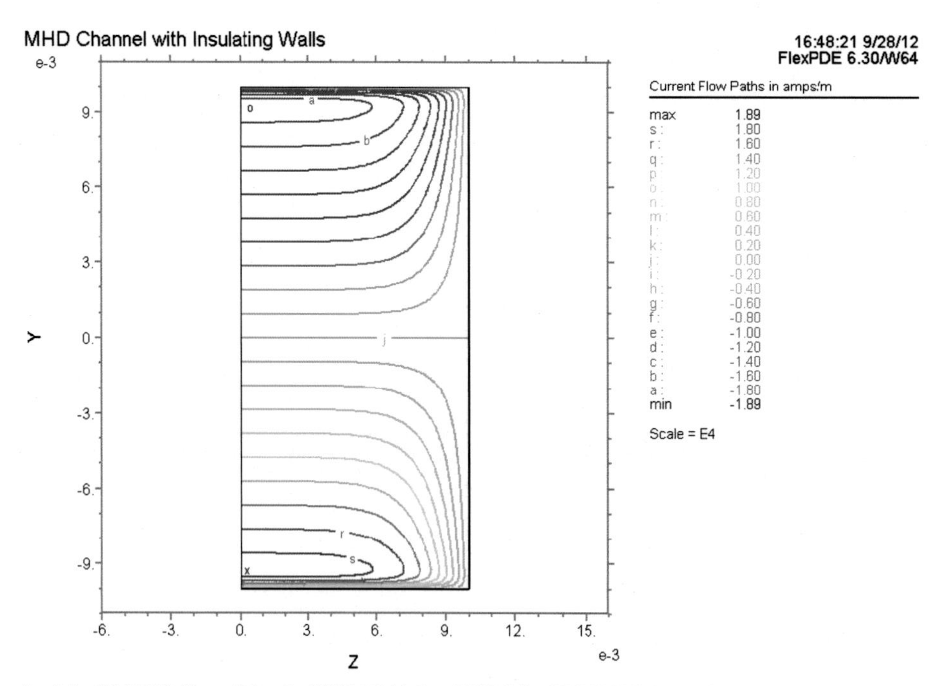

Figure 68: Magnetic intensity for M = 100

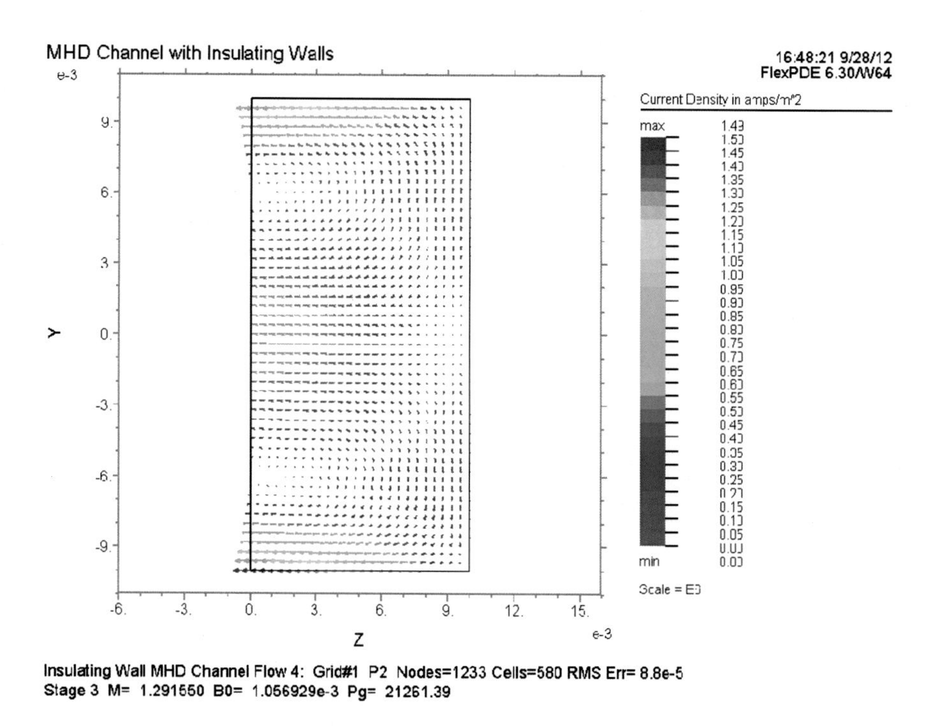

Figure 69: Current vectors for M = 1.29155

accelerating force is counteracted by the near the boundary thus meeting the usual no slip . These calculations could have been done from the closed form expressions given here and would have been a more difficult task. In addition changes in the geometry, for example to a channel of irregular cross section would have been virtually impossible whereas they are easily executed by use of the method of finite elements. In Fig.73 and Fig.74 note how the induced magnetic force changes faster near the walls as the Hartmann number increases.

9.6 Magnetohydrodynamic problems for reader solution

9.6.1 Sea water flowing in an insulating pipe

A 23 cm inner diameter pipe with insulating walls has seawater flowing through it and there is a large magnetic field applied perpendicular to the flow. The pressure gradient should be regulated so that the Reynolds number is 2500 in order to avoid turbulence. Calculate the emf generated as the Hartmann number increases from zero to 5. Hint: Because of the physical properties of sea water are so much different that lithium be sure to adjust the flow so that Re $=2500$ at all values of the Hartmann number. See Appendix I for the solution.

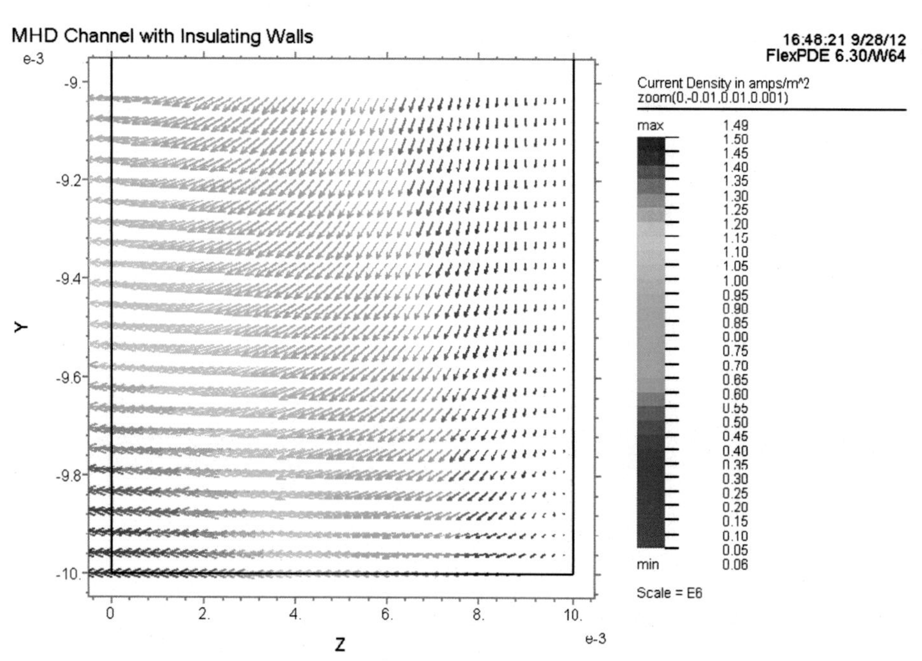

Figure 70: Zoomed current vectors for M = 1.2916

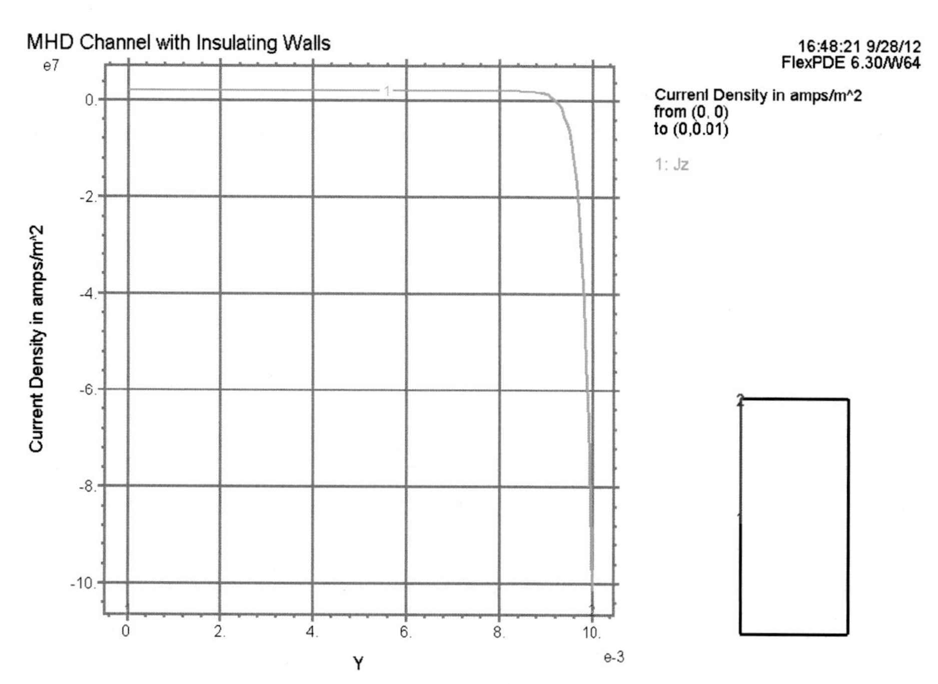

Figure 71: J_z in the channel center for M = 100

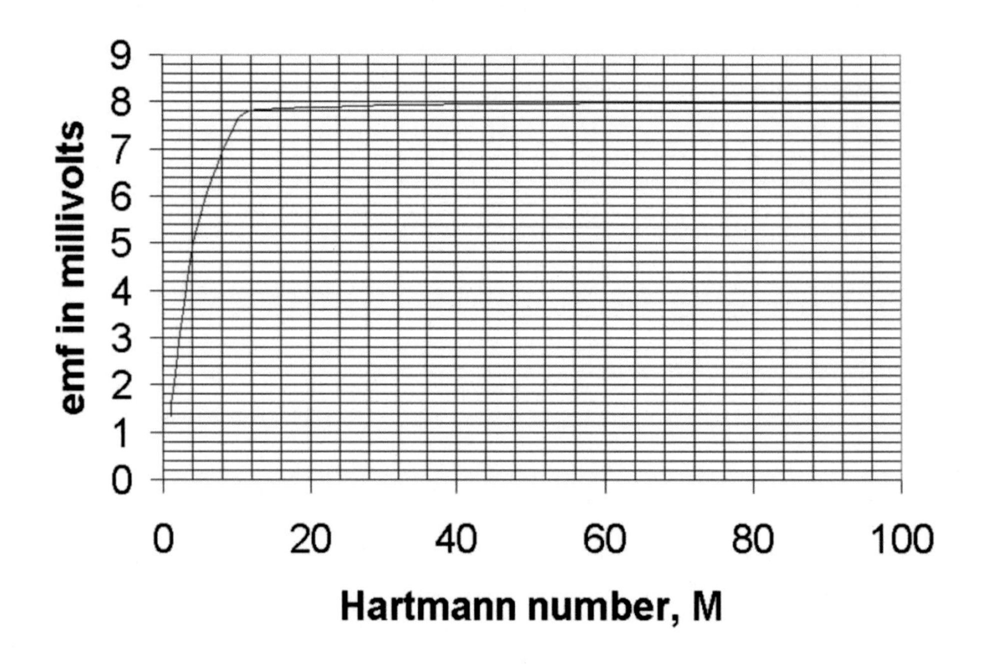

Figure 72: Emf vs. M for constant channel velocity

9.6.2 Sea water flowing in an ideally conducting wall pipe

Use the same diameter and Hartmann numbers as in the previous problem. Hint: In this problem the use of symmetry that was so helpful in the previous problem may lead to instabilities making convergence unlikely. Here it is best to use the geometry of the entire pipe cross section. See Appendix J for the solution.

9.6.3 Do the geometry of 9.5.3 with ideally conducting walls

The pressure gradient should be regulated so that the Reynolds number is 2500 in order to avoid turbulence. Calculate the emf generated as the Hartmann number increases from zero to 100. See Appendix K for the solution.

9.6.4 Half full pipe of 9.6.1

(a) Consider the insulating wall pipe when it is half full of flowing sea water. Calculate the emf generated as the Hartmann number increases from zero to 5. Hint: At the interface between air and sea water $\partial u/dn = 0$. See Appendix L for the solution.

(b) Solve the same problem when the pipe wall is an ideal conductor. See Appendix M for a hint and results.

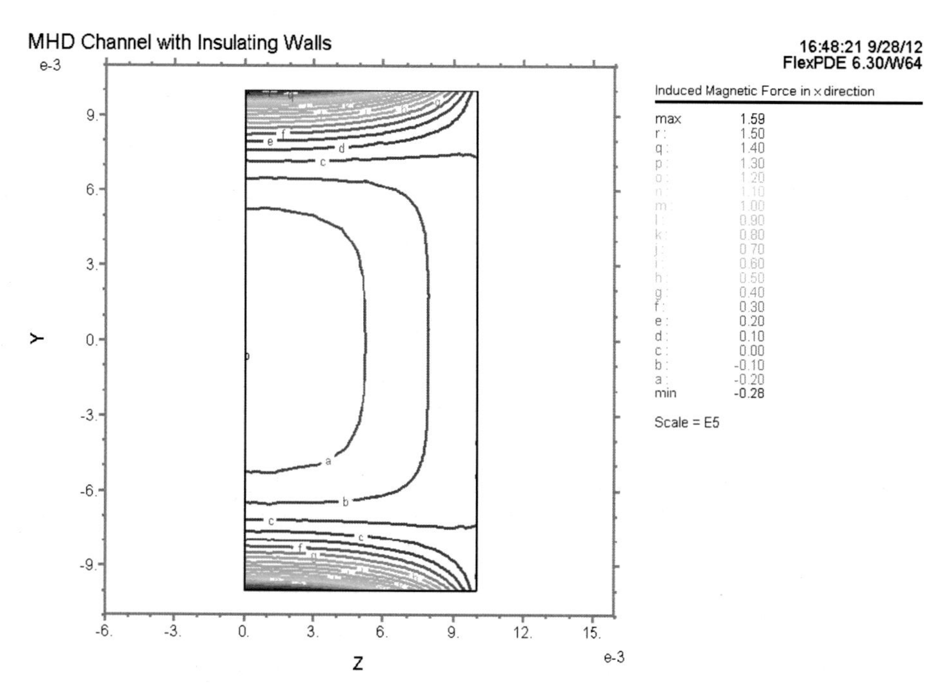

Figure 73: Induced magnetic force for M = 12.92

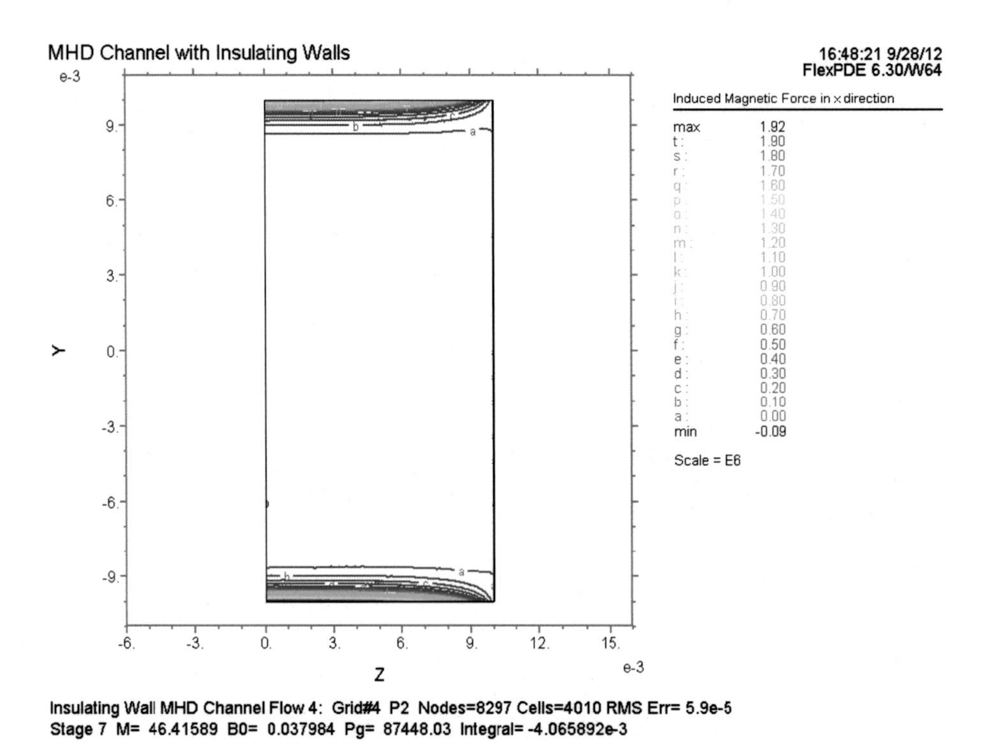

Figure 74: Induced magnetic force for M = 46.42

9.6.5 Half full rectangular channel

(a) Consider the geometry of 9.5.3 when it is half full of liquid lithium and has perfectly insulating walls. Find the emf.

(b) Redo this problem with ideally conducting wall. Find the emf.

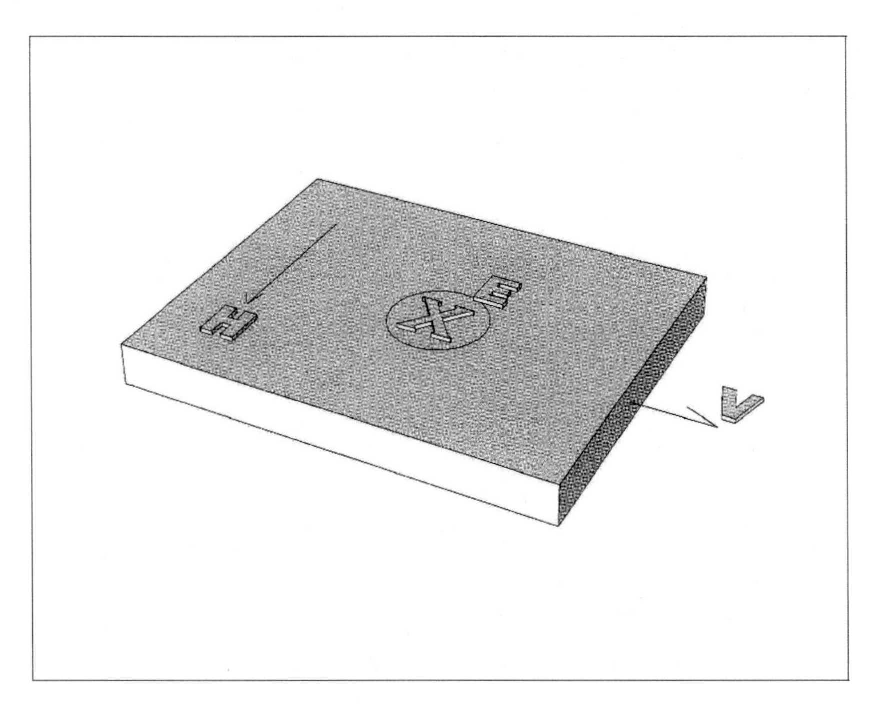

Figure 75: Glass plate moving through a magnetic field

9.7 A pane of glass moving through a magnetic field

To illustrate the application of these transformation equations, consider the case of a large pane of in a plane parallel to a magnet field intensity, **H** and moving perpendicularly with velocity **V** through this field in the plane of the glass as shown in Fig. 75. There is no free charge on the glass surfaces. The problem is to determine the fields seen both by an observer stationary with respect to the magnet producing the field (unprimed quantities) and by an observer on the glass plate (primed quantities). Let the subscripts 0 and ℓ denote free space and glass, respectively. The given conditions are: $H = H_0$ and $E_0 = 0$. The known material constitutive equations are $\mathbf{B}_0 = \mu_0\mathbf{H}_0, \mathbf{B}_\ell^{'} = \mu_0\mathbf{H}_\ell^{'}, \mathbf{B}_0^{'} = \mu_0\mathbf{H}_0^{'}, \mathbf{D}_0 = \mu_0\mathbf{E}_0, \mathbf{D}_\ell^{'} = \mu_0\mathbf{E}_\ell^{'}$ and $\mathbf{D}_0^{'} = \mu_0\mathbf{E}_0^{'}$. From Eq.64

$$\mathbf{D}_{0_\perp}^{'} = \beta[\mathbf{D}_0 + (\mathbf{V} \times \mathbf{H}_0)/c^2]_\perp \simeq \epsilon_0 V B_0 \tag{150}$$

where it is assumed that $V/c \ll 1$ and neglecting the stretcher operator. Similarly by Eq.65

$$\mathbf{H}_{0_\perp}^{'} = \beta(\mathbf{H}_0 - \mathbf{V} \times \mathbf{D}_0)_\perp = \beta\mathbf{H}_0 \simeq \mathbf{H}_0 \tag{151}$$

The only fields that are unknown are $\mathbf{E}_\ell$ and $\mathbf{B}_\ell$ which can be found by employing the inverse relationships of Eq.63 and Eq.66. The inverse transformations are found by interchanging primed and unprimed quantities and by replacing

$\mathbf{V}$ by $-\mathbf{V}$. Making the same assumptions as before we obtain

$$E_\ell = (\epsilon_0/\epsilon_\ell)VB_0 = -(1-\epsilon_0/\epsilon_\ell)VB_0 \tag{152}$$

and

$$B_\ell = B_0 - (\epsilon_0/\epsilon_\ell)\,(V/c)^2\, B_0 \simeq B_0 \tag{153}$$

These results can be verified by the use of the relative constitutive relations given by Eq.80 and Eq.82. The results are tabulated in the following table:

	$\Im_0$	$\Im_0'$	$\Im_\ell$	$\Im_\ell'$
E	0	VB_0	$-(1-\epsilon_0/\epsilon_\ell)VB_0$	$(\epsilon_0/\epsilon')VB_0$
D	0	$\epsilon_0 VB_0$	0	$\epsilon_0 VB_0$
H	H_0	H_0	H_0	H_0
B	B_0	B_0	B_0	B_0

where $\Im$ represents the fields given in the first column. It is noteworthy that even in this simple case, the "*motional field*" $\mathbf{V}\times\mathbf{B}$ does not polarize the dielectric in the same way as does a "*static field*". In addition the electric field in the glass is not only different in magnitude for the different observers but also *differs in sign.*

9.8 Current-carrying wire in relative motion

A wire carrying a direct current $i'$is shown in Fig.76. To an observer perched on the wire the current in the wire is i' and the magnetic induction outside the wire is $\mathbf{B}' = \boldsymbol{\phi}(\mu_0 i'/2\pi r)$. What fields would and observer traveling in the -x'direction with a speed of V_0 see? Would he think the current in the wire differs from i'? Note: $\boldsymbol{\phi}$, $\mathbf{r}$ and $\mathbf{x}$ are unit vectors in their respective directions.

9.8.1 Solution

The fields seen by the moving observer are given by the MLT. In the air, from the inverse of Eq.63 $\mathbf{E}_\perp = \beta(\mathbf{E}' - \mathbf{V}\times\mathbf{B}')_\perp$. In this case $\mathbf{E}_\perp' = 0$ and $\mathbf{V}\times\mathbf{B}_\perp' = -\mathbf{r}(\beta\mu_0 V_0 i'/2\pi r)$. Hence

$$\mathbf{E}_\perp = \mathbf{r}(\beta\mu_0 V_0 i'/2\pi r) \text{ and } \mathbf{E}_\parallel = \mathbf{x} i'/\sigma A \tag{154}$$

where A is the cross sectional area of the wire. From the inverse of Eq.66

$$\mathbf{B}_\perp = \beta B_\perp' = \boldsymbol{\phi}\beta(\mu_0 i'/2\pi r) \text{ and } \mathbf{B}_\parallel = 0 \tag{155}$$

The $\mathbf{D}$ and $\mathbf{H}$ fields are no problem because the same constitutive equations hold for either observer in free space. The $\mathbf{D} = \epsilon_0\mathbf{E}$ and $\mathbf{H} = \mathbf{B}/\mu_0$. From Eq.154 it is clear that the moving observer sees a radial electric field not seen by an observer on the wire. In accordance with Eq.155 the moving observer sees a circumferential magnetic induction that is modified by the factor β. The moving observer might attribute the electric field to a net charge accumulated

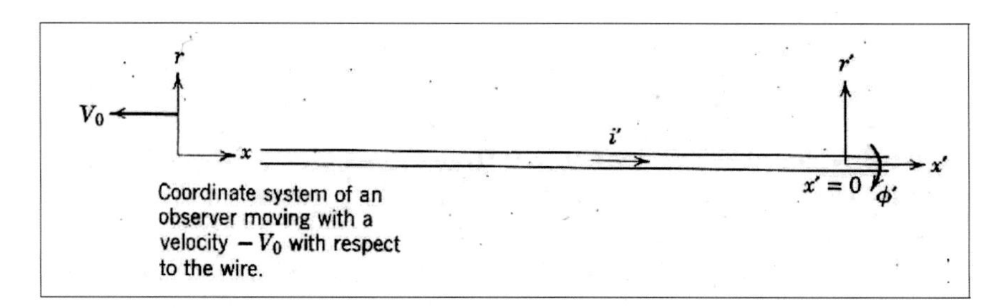

Figure 76: A coordinate system moving with respect to a wire

in and/or on the wire provided he did not know it was moving with respect to him. This would be given by

$$\rho_w = \beta V_0 i' / Ac^2 \tag{156}$$

in coulombs per meter of wire and he would think the current in the wire is $i = \beta i'$. At velocities usually encountered $\beta \backsimeq 1$; thus the most interesting result occurring at velocities much less than the speed of light is the observation of charge on the wire seen by the moving observer. It is interesting to find the fields inside the wire, also. To the observer on the wire the electric field in the wire is

$$\mathbf{E}'_w = \mathbf{x} i' / \sigma A \tag{157}$$

where σ is the electrical conductivity of the wire. By Ampere's circuital law, $\oint \mathbf{H} \cdot d\ell = \int_S \mathbf{J} \cdot d\mathbf{S}$ the magnetic induction or flux density in the wire is given by

$$\mathbf{B}'_w = \phi \mu_w i' r / 2A$$

Here μ_w is the wire permeability in the rest frame. The observer moving in the $-x'$ direction with speed V_0 sees an electric field given from Eq.71 given by

$$\mathbf{E}_w = \mathbf{r} \beta \mu_w V_0 i' r / 2A + \mathbf{x} i' / \sigma A \tag{158}$$

and there results from Eq.72

$$\mathbf{B}_w = \phi \mu_w i' r / 2A \tag{159}$$

In addition, the MLT may be applied to the current density and charge in the wire. By Eq.73

$$\mathbf{J}_w = x' \beta \mathbf{J}'_w = \mathbf{x} \beta i' / A \tag{160}$$

and because $\rho' = 0$ by Eq.69

$$\rho_w = \beta V_0 i' / c^2 A \tag{161}$$

Eq.160 is in agreement with current in the wire seen by the moving observer. To check Eq.161, the equation for the space charge seen in the wire by the moving observer, D_{rw} is calculated from Eq.161 and compared to D_{rw} computed from the constitutive Eq.80. To calculate D_{rw} Gauss's Law is used as: $D_{rw}2\pi r\text{l}=\beta V_0 i^{'}\pi r^2\text{l}/c^2A$ where l is the length of the wire in the $x^{'}$direction, whence

$$D_{rw} = \beta V_0 i^{'}\pi r/2c^2A \qquad (162)$$

The substitution of $\mathbf{E}_w$ from Eq.158 and $\mathbf{B}_w$ from Eq.159 into Eq.79 yields the same value of D_{rw} . In summary, the fields seen by the observer are given in the following table:

	Space	Wire
D	$\mathbf{r}\beta i^{'}V_0/2\pi c^2r + \mathbf{x}\epsilon_0 i^{'}/\sigma A$	$\mathbf{r}\beta i^{'}rV_0/2c^2A + \mathbf{x}\epsilon_w i^{'}/\sigma A$
E	$\mathbf{r}\beta\mu_0 i^{'}V_0/2\pi r + \mathbf{x}i^{'}/\sigma A$	$\mathbf{r}\beta\mu_w i^{'}rV_0/2A + \mathbf{x}i^{'}/\sigma A$
B	$\phi\beta\mu_0 i^{'}/2\pi r$	$\phi\beta\mu_w i^{'}/2A$
H	$\phi\beta i^{'}/2\pi r$	$\phi\beta i^{'}/2A$
J	0	$\mathbf{x}\beta i^{'}/A$
ρ	0	$V_0\beta i^{'}/c^2A$

Those seen by the observer in the rest frame of the wire are presented in the next table.

	Space	Wire
$\mathbf{D}^{'}$	$\mathbf{x}\epsilon_0 i^{'}/\sigma A$	$\mathbf{x}\epsilon_w i^{'}/\sigma A$
$\mathbf{E}^{'}$	$\mathbf{x}i^{'}/\sigma A$	$\mathbf{x}i^{'}/\sigma A$
$\mathbf{B}^{'}$	$\phi\mu_0 i^{'}/2\pi r$	$\phi\mu_w i^{'}r/2A$
$\mathbf{H}^{'}$	$\phi i^{'}/2\pi r$	$\phi i^{'}r/2A$
$\mathbf{J}^{'}$	0	$i^{'}/A$
$\rho^{'}$	0	0

To the moving observer the wire has an anisotropic permittivity and conductivity. It contains a positive space charge per unit length of magnitude $\beta V_0 i^{'}/c^2$. When $\beta \to 1$, commonly called the magnethydrodynamic approximation, although the moving and the rest frame observers see different values of electric field intensity, **E** and electric displacement, **D** they see the same values of magnetic intensity, magnetic induction and current density. Any differences in the latter quantities might be called high velocity relativistic effects, whereas the other differences, including the appearance of ρ_w are low velocity relativistic effects.

9.9 Glass plate generator

In Fig.77 is a thin glass plate of sufficient width and breadth as to justify calling it an infinite sheet located in the xz plane. It moves in the x direction with velocity V through a magnetic intensity field H_{0z} . The problem is to find the emf induced by the motion of the glass plate. We calculate the electric field seen by an observer riding on the plate, find the terminal voltages for different observers from the electric fields and than evaluate $\int \mathbf{E}^{'}\, d\ell^{'}$ to find the potential drop in the moving dielectric. Then it is shown that both observers see the same value of terminal voltage at glass plate speeds likely to be encountered in a terrestrial laboratory. Let the subscripts 0 and 1 denote free-space and glass-plate variables, respectively. By the continuity of the tangential components of magnetic field in the rest frame of the glass plate there results

$$H_{1z} = H_{0z} \tag{163}$$

Assuming that no surface charge (real, not polarization charge) rests on the surface of the glass, the normal component of the electric induction is continuous, also. Thus

$$D'_{1y} = D'_{0y} \tag{164}$$

and from Eq.150 because $D_{0y} = 0$

$$D'_{0y} = -\beta H_0 V/c^2 = -\beta \epsilon_0 B_0 V \tag{165}$$

Therefore,

$$E'_{1y} = D'_{1y}/\epsilon = -\beta B_0 V \epsilon_0/\epsilon \tag{166}$$

and because the normal components of $\mathbf{D}$ is continuous and

$$E'_{0y} = -\beta B_0 V \tag{167}$$

Remembering that the filaments connecting the brushes to A and B are not conductors, the rise in potential from A to B is calculated making use of Eq.71 to obtain $\mathbf{E}'$. The following steps result in Φ'_T:

$$\Phi'_T = -\int_A^B \mathbf{E}' \cdot d\ell = -\int_{d/2}^{-d/2} [(-\beta B_0 V \epsilon_0/\epsilon) + (\beta B_0 V)]dy = \beta B_0 V d(1 - \epsilon_0/\epsilon) \tag{168}$$

It is interesting to note that Eq.150 yields $D_{1y} = 0$. Here we note that the electric induction $\mathbf{D}$ is differs to different observers. From inverse of Eq.71

$$E_{1y} = \beta(E'_{1y} + VB'_{1z}) = \beta^2 B_0 V(1 - \epsilon_0/\epsilon) \tag{169}$$

According to an observer riding on the external circuit, the rise in potential from A to B is

$$\Phi_T = \int_{-d/2}^{d/2} E_{1y} dy = \beta^2 B_0 V(1 - \epsilon_0/\epsilon) \tag{170}$$

Thus, $\Phi_T = \beta \Phi'_T$, indicating that only at very high velocities do different observers see different values of potential. The emf can be calculated by Eq.19 and assuming $\beta \cong 1$, then

$$e = \oint \mathbf{v} \times \mathbf{B} \cdot dl \tag{171}$$

where the observer is stationary with respect to the external circuit. The emf is $e = VB_0 d$. The terminal voltage is given by

$$V_T = e - \int_A^B \mathbf{E}'_1 \cdot dl' = VB_0 d(1 - \epsilon_0/\epsilon) \tag{172}$$

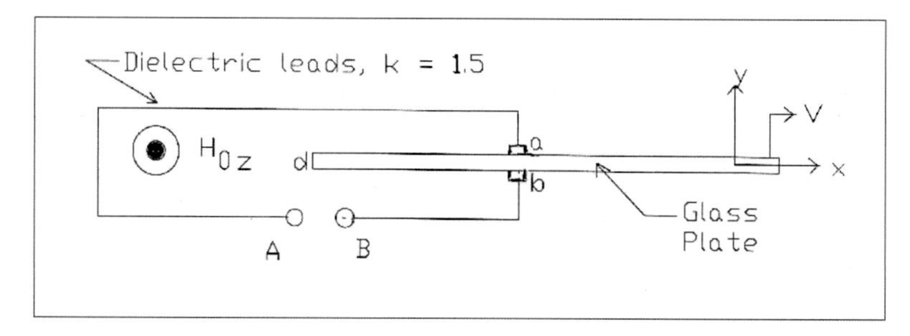

Figure 77: Glass plate moving in a magnetic field

which is the same as the potentials previously calculated provided $\beta \cong 1$. This is no cause for alarm, because Eq.19 from which Eq.172 is derived, is based upon the assumption that $\beta = 1$.

In this example, the internal potential drop in the generator cannot be neglected, as the generator is a dielectric rather than a conductor. it is interesting to note that the motional field $\mathbf{V} \times \mathbf{B}$ does not polarize the dielectric in the same way as would a static electric field.

9.9.1 Field solution

It is interesting to solve the field problem posed by the geometry just considered here. Let the glass plate relative permittivity be 5 whilst the leads have 1.5 and the surrounding air 1. In Fig.78 is depicted the geometry in which the field solution has been executed.All dimensions are in meters. Fig.79 exhibits the electric field vectors resulting when the plate moves to the right through a magnetic induction field of 1 Tesla at a speed of 1 meter/second. The glass plate is 10 centimeters thick and the leads are 1 cm thick. In Fig.80 is shown the second quadrant of Fig.79 to get a better view of the electric intensity vectors.

In Fig.81 is shown a plot of the y-component of the electric intensity between points a and b in Fig.77. It is about -0.99 compared to the -1 indicated by the simple analysis given above. From this figure $\int_b^a \mathbf{E}_1^{'} \cdot dl^{'} = -0.0987$. Fig.82 shows the variation of E_y along the center line of the glass plate. This plot indicates that end effects are not very important unless $x \prec -0.4$ or $x \succ 0.4$. A complete three dimensional solution yields almost identical results.

9.10 A TEM transmission line in relative motion

In several examples observers in relative motion with respect to conductors and other material media see different fields from those seen in the rest frame of conductors and other material media. In this example the coaxial TEM transmission line carrying alternating current is considered. We want to determine how the fields appear to an observer moving in the -z direction (see Fig.83) with

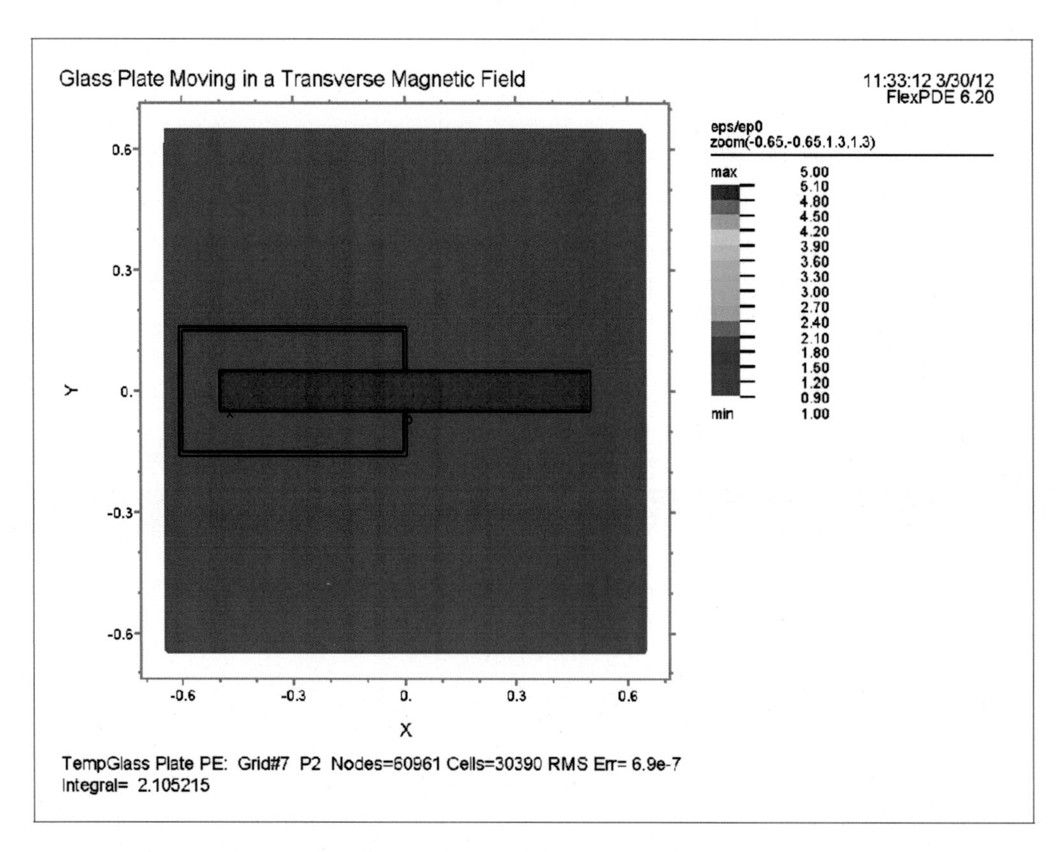

Figure 78: Glass plate and leads geometry

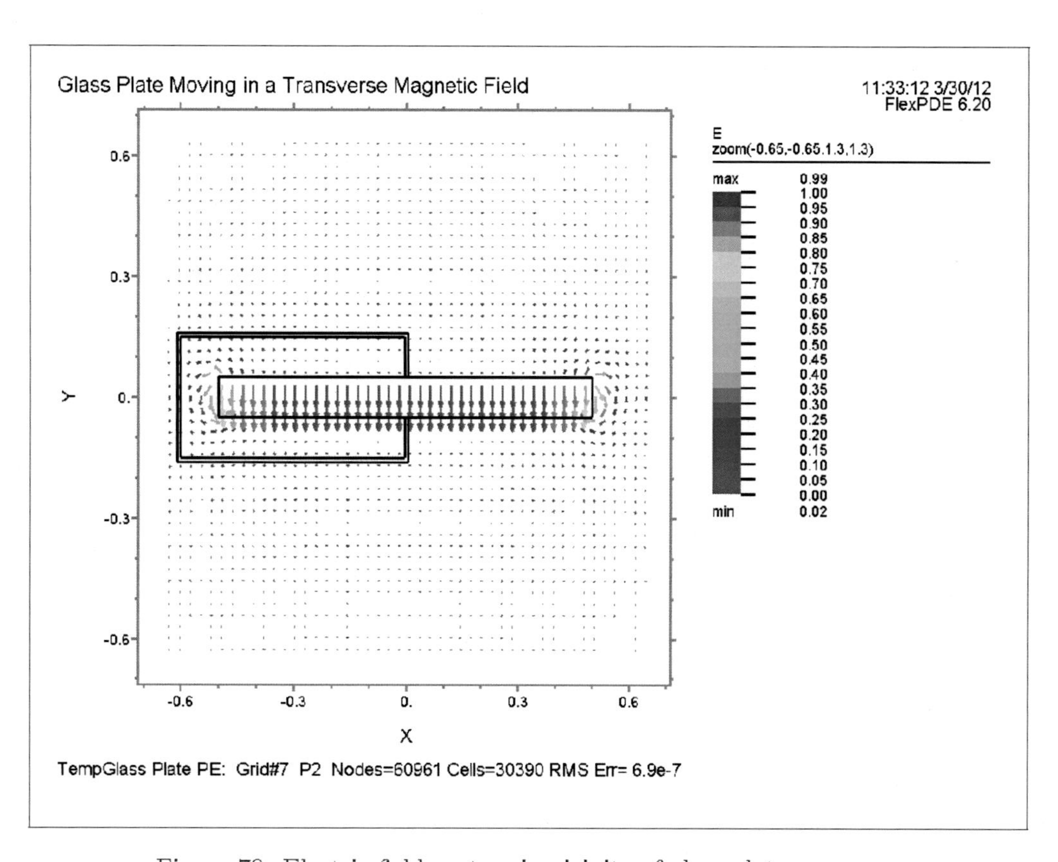

Figure 79: Electric field vectors in vicinity of glass plate

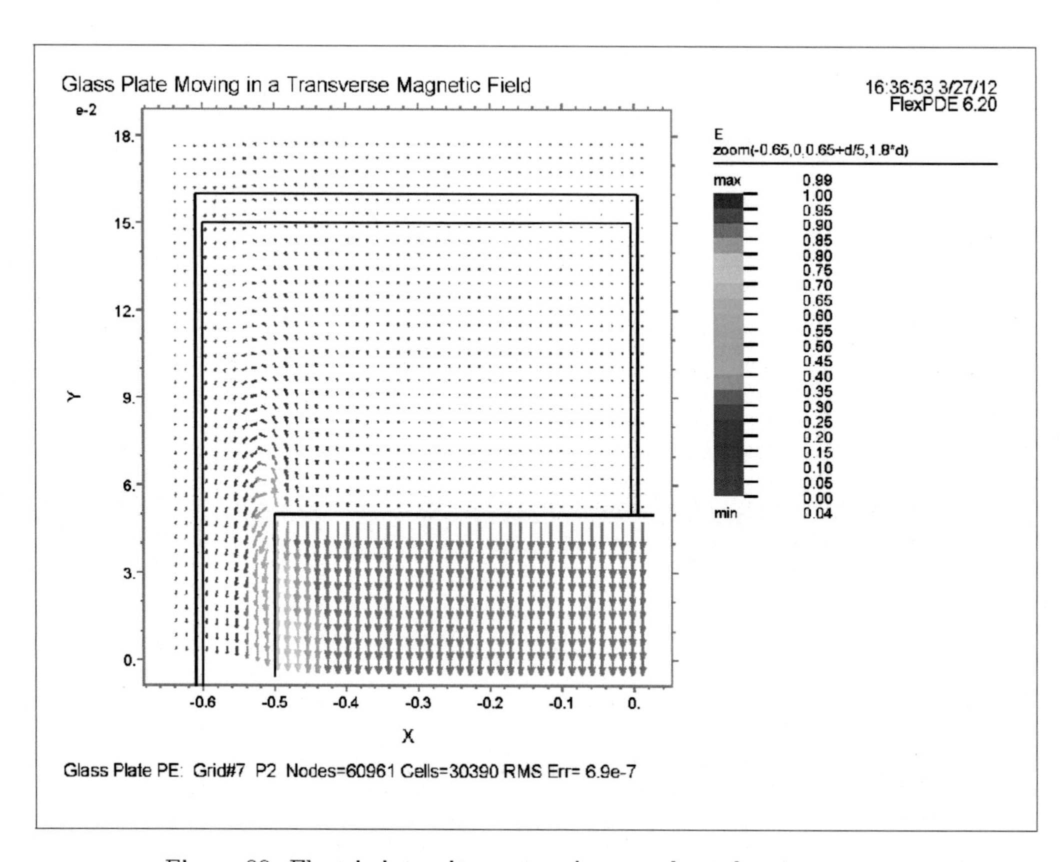

Figure 80: Electric intensity vectors in second quadrant

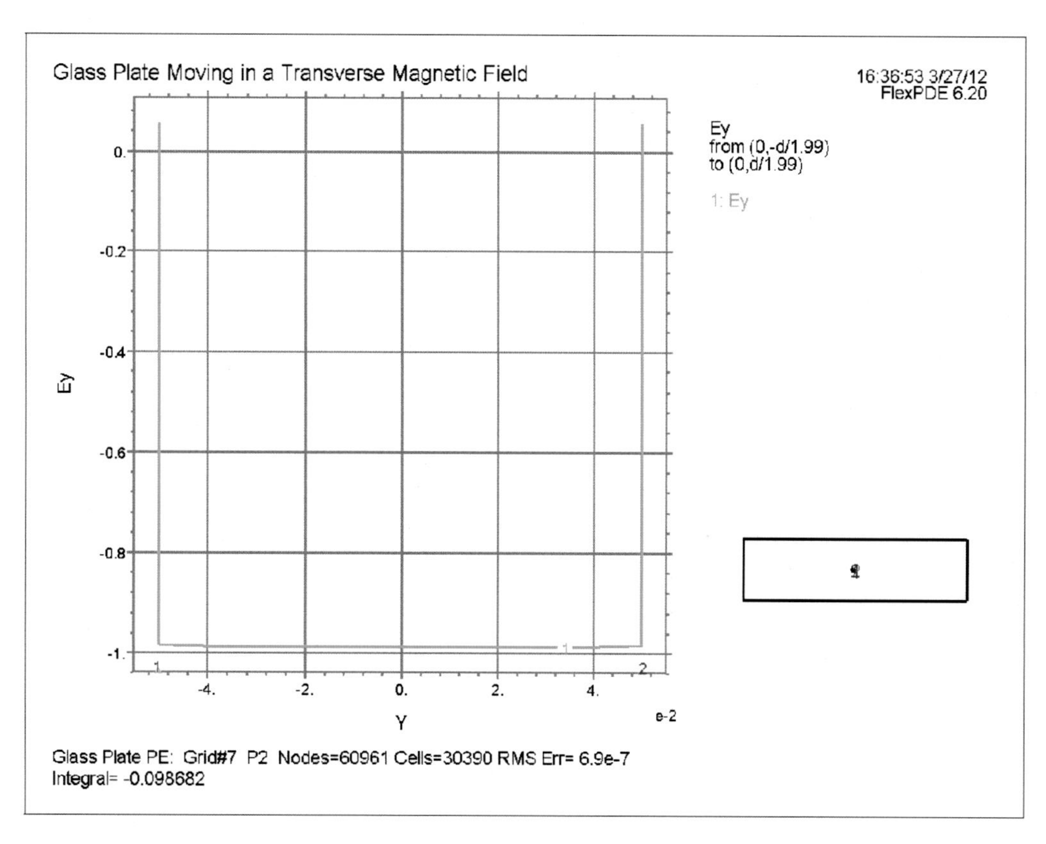

Figure 81: E_y across the plate thickness

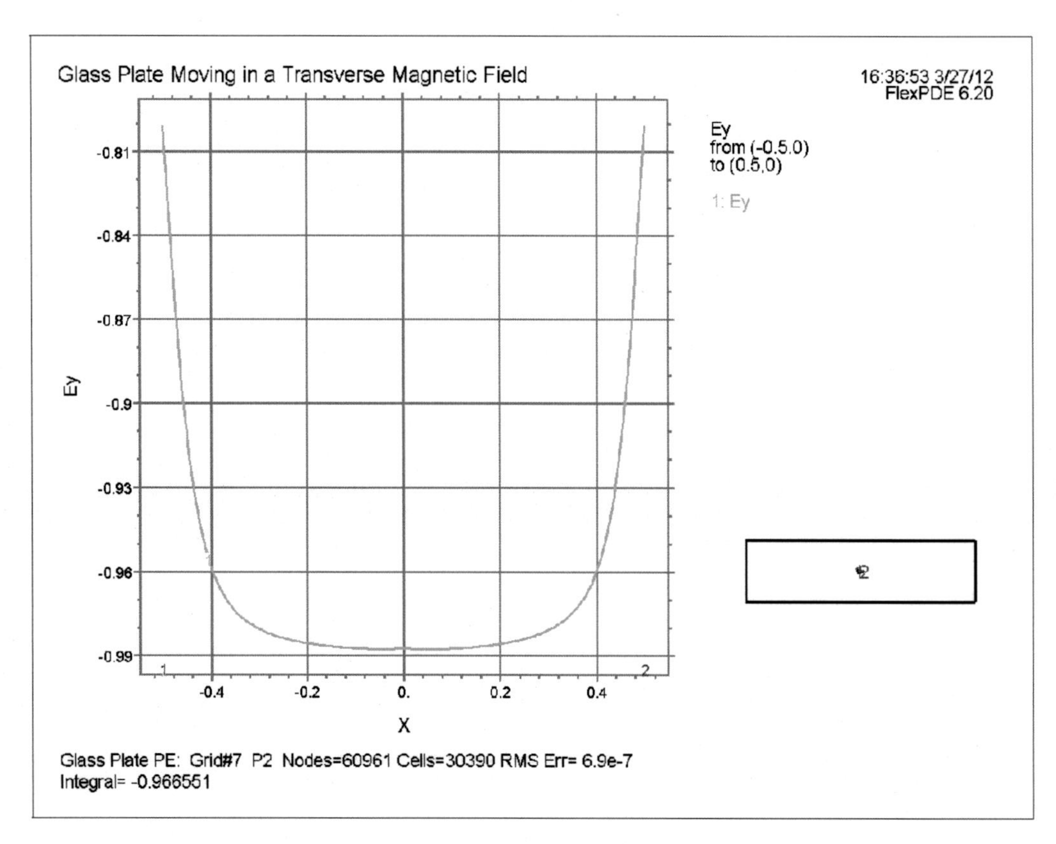

Figure 82: Plot of E_y along the glass plate center line

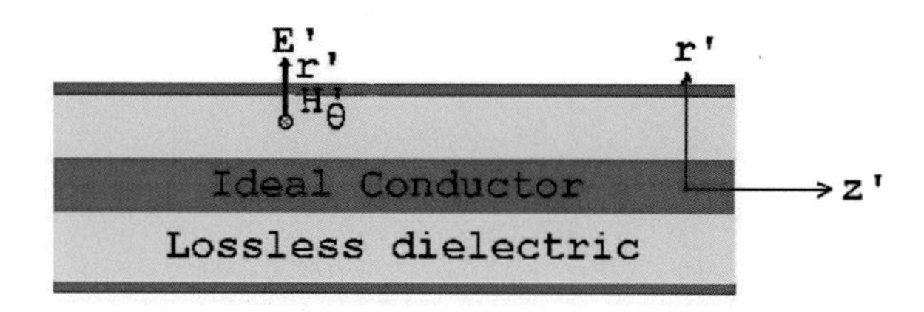

Figure 83: A coaxial transmission line

speed of $V_0$with respect to the transmission line.

First the fields must be found in the rest frame. Assuming a transverse electromagnetic mode, there will be only two field components, $E_r^{'}$ and $H_\theta^{'}$. Here phasor variations of the form $F^{'}(r^{'},z^{'},\theta^{'})e^{j\omega^{'}t^{'}}$ are assumed to represent cisoidal time variation. Then the two Maxwell curl equations become

$$dE_{r'}^{'}/dz^{'} = -j\omega^{'}\mu H_{\theta'}^{'} \tag{173}$$

$$dH_{\theta'}^{'}/dz^{'} = -j\omega^{'}\epsilon E_{r'}^{'} \tag{174}$$

It is important to realize that Maxwell's equations, when written in any medium except free space, are simplest in the rest frame; otherwise they become tangled by the constitutive equations. Combining these equations yields

$$d^2/dz^{'2}\begin{vmatrix}\mathbf{E}_{r'}^{'}\\ \mathbf{H}_{\theta'}^{'}\end{vmatrix} + \omega^{'2}\mu\epsilon\begin{vmatrix}\mathbf{E}_{r'}^{'}\\ \mathbf{H}_{\theta'}^{'}\end{vmatrix} = 0 \tag{175}$$

The solution to Eq.175 is

$$\begin{vmatrix}\mathbf{E}_{r'}^{'}\\ \mathbf{H}_{\theta'}^{'}\end{vmatrix} = E_0^{'}\begin{vmatrix}\mathbf{1}\\ (\epsilon/\mu)\end{vmatrix}e^{-jz^{'}\omega^{'}(\mu\epsilon)^{1/2}} \tag{176}$$

where $E_0^{'}$ is a constant and $(\mu/\epsilon)^{1/2} = Z_m$ the intrinsic impedance of the dielectric as seen by an observer in the rest frame. Before transforming the electric and magnetic field intensities into quantities seen by the moving observer, time must be introduced into Eq.176. It then becomes

$$\begin{vmatrix} \mathbf{E}'_{r'}(z',t') \\ \mathbf{H}'_{\theta'}(z',t') \end{vmatrix} = E'_0 \begin{vmatrix} \mathbf{1} \\ (\epsilon/\mu) \end{vmatrix} e^{-j\omega'[t'-z'(\mu\epsilon)^{1/2}]} \tag{177}$$

To determine what the moving observer sees, all the primed quantities in this equation must be transformed by the Lorentz and the Maxwell-Lorentz transformations. By Eq. 66 there results

$$\begin{vmatrix} E_0 \\ H_0 \end{vmatrix} = \beta \begin{vmatrix} \mathbf{1} \\ (\epsilon/\mu) \end{vmatrix} E'_0 \tag{178}$$

It is not really necessary to transform the magnitude of the $H'_{\theta'}$ field because Eq.177 relates electric and magnetic intensity. The quantity $\omega'[t' - z'(\mu\epsilon)^{1/2}]$ is transformed the use of

$$z' = \beta(z - Vt) \tag{179}$$

and

$$t' = \beta(t - Vz/c^2) \tag{180}$$

to yield

$$\omega'[t' - z'(\mu\epsilon)^{1/2}] = \beta\omega'\{[1 + V_0(\mu\epsilon)^{1/2}]t - [(V_0/c)^2 + V_0(\mu\epsilon)^{1/2}]z/V_0\} \tag{181}$$

Because of the longitudinal Doppler shift

$$\omega = \beta\omega'[1 + V_0(\mu\epsilon)^{1/2}] = \beta\omega'(1 + V_0/c_m) \tag{182}$$

where c_m is the speed of light in the transmission line dielectric in the rest frame. The transformed fields seen by the moving observer become

$$E_r(z,t) = \beta(1 + V_0/c_m)E'_0 e^{j\omega\{t - [V_0^2/c^2 + V_0/c_m]/[V_0(1+V_0/c_m)]z\}} \tag{183}$$

and

$$H_\theta = \beta/Z_m \tag{184}$$

By the use of Eq.74 D_r and B_θ are found. The results are given by

$$D_r(z,t) = \beta\epsilon(1 + V_0 c_m/c^2)E'_0 e^{j\omega\{t - [V_0^2/c^2 + V_0/c_m]/[V_0(1+V_0/c_m)]z\}} \tag{185}$$

and

$$B_\theta(z,t) = (\beta/c_m)(1 + V_0 c_m/c^2)E'_0 e^{j\omega\{t - [V_0^2/c^2 + V_0/c_m]/[V_0(1+V_0/c_m)]z\}} \tag{186}$$

Before examining the physical significance of the magnitudes of the fields seen by the moving observer, we should note that in addition to seeing the Doppler shift in frequency, he sees a different phase velocity, V_p, This velocity, which is the reciprocal of the coefficient of z, is

$$V_p = (1 + V_0/c_m)/(V_0/c^2 + 1/c_m) \tag{187}$$

Because the phase velocity to the observer in the rest frame is $V_p^{'} = c_m$ Eq.187 becomes

$$V_p = (V_0 + V_p^{'})/(1 + V_0 V_p^{'}/c^2) \tag{188}$$

which is true for all values of $V_0 \prec c$. When $V_0^2 \ll c^2$ Eq.188 simplifies by observing that $1/(1 + V_0 V_p^{'}/c^2) \approx 1 - V_0 V_p^{'}/c^2$ to

$$V_p \approx V_p^{'} + V_0(1 - V_p^{'}/c^2) \tag{189}$$

The term $1 - V_p^{'}/c^2$ is known as the Fresnel dragging coefficient . Nonrelativistic theories such as the Lorentz electron theory yielded this result only when the medium was nonmagnetic. The form of the Fresnel dragging coefficient derived here was verified by M. and H. A. Wilson[5] in a 1913 experiment suggested by Einstein and Laub . The examination of Eq.183 and Eq. 115 indicates that for $V_0^2 \ll c^2$ the moving observer sees a significantly different magnitude for the E and H fields than the observer stationed on the medium. To both observers the D and B fields are the same.

Inspection of the aforementioned equations shows that E_r and H_θ both equal zero if $V_0 = -V_p^{'}$ with speed c_m. This produces an interesting limiting case for checking the results derived herein. Since the observer is traveling at the phase velocity in the direction of electromagnetic wave propagation, we would expect him to say $\omega = 0$. this, indeed, is a result of this analysis. But why does this observer see electric displacement and magnetic induction but no electric or magnetic intensity? To those unaccustomed to problems of the electrodynamics of moving media this could be explained only by infinite permittivity and permeability. However, the constitutive equations Eq.84 and Eq.85 indicate that, although E and H are zero, D and B need not be zero and they yield $D_r = P_r$ and $B_\theta = \mu_0 M_\theta$ when $\mathbf{V}_0 = -\mathbf{z}V_p^{'}$.

10 More advanced problems solution

10.1 Finite conductivity for glass plate of Example 9.6

If the glass plate considered in Example 9.6 had finite conductivity, what would the steady state charge density on the surface be?

10.1.1 Glass plate covered by conducting metal

The two large surfaces of the glass plate are plated with metal and shorted with a jumper stationary with respect to the magnet as shown in Fig.84. (a) What are the fields seen by observers on the glass plate and the jumper? (b)What total charge would flow in the circuit just after the jumper has touched the metal on both sides of the glass plate?.

10.2 Thin traveling in a magnetic field

Work Example 10.1, for a thin glass rod as depicted in Fig.85.

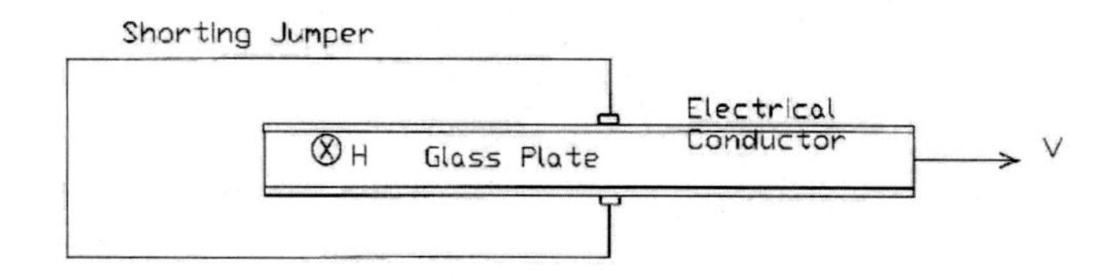

Figure 84: Shorted conductor plated glass pane

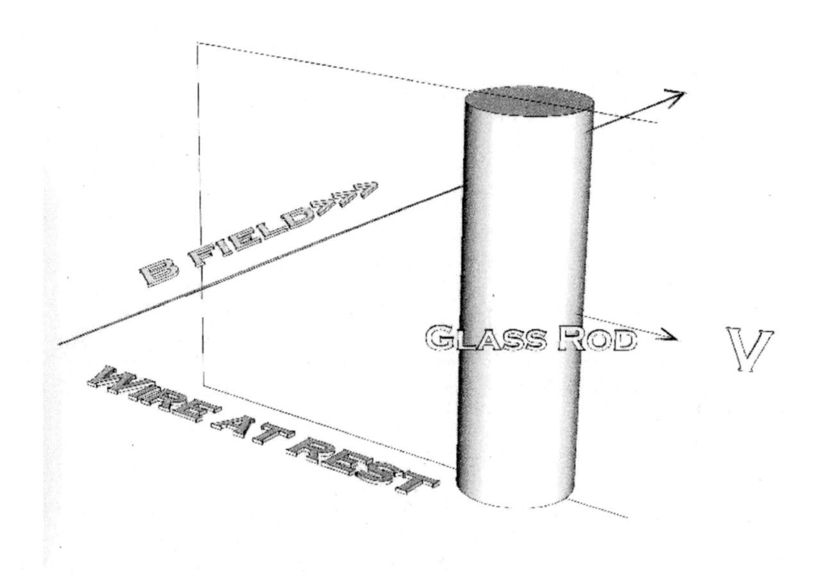

Figure 85: Glass rod traveling in magnetic field

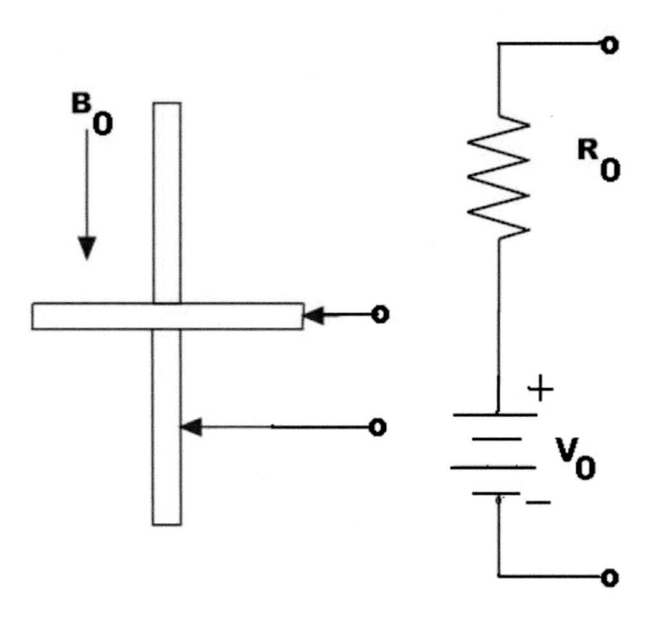

Figure 86: Equivalent circuit of Faraday disc

10.3 Thin glass rod traveling very fast in a magnetic field

Rework Problem 11.2 without making the assumption that the glass rod's velocity is not much less than the speed of light.

10.4 Equivalent circuits of the Faraday disc generator

If a Faraday disc generator is made of metal, the equivalent circuit is merely a battery V_0 in series with a resistance R_0 where R_0 is the radial resistance of the disc. assuming a perfectly conducting rim. If there is no armature reaction V_0 is given by

$$V_0 = B_0\omega(a^2 - b^2) \tag{190}$$

where B_0 is the constant applied magnetic induction as shown in Fig.86 Here a and b are the outer and inner radii of the disc, respectively. The internal resistance is given by

$$R_0 = (1/\sigma)\int_b^a dr/(2\pi r) = (2\pi\sigma)^{-1}\ln(a/b)$$

Show that if the metal disc is replaced by a disc of the same size made of a leaky, ferromagnetic dielectric with the inside and outside cylindrical surfaces copper plated, the equivalent circuit is given by Fig.87 where V_0 is the same as for the metal disc of Fig.86. C is the capacitance of the disc when considered a simple capacitor and R_1 is the resistance between the inner and outer surfaces of the

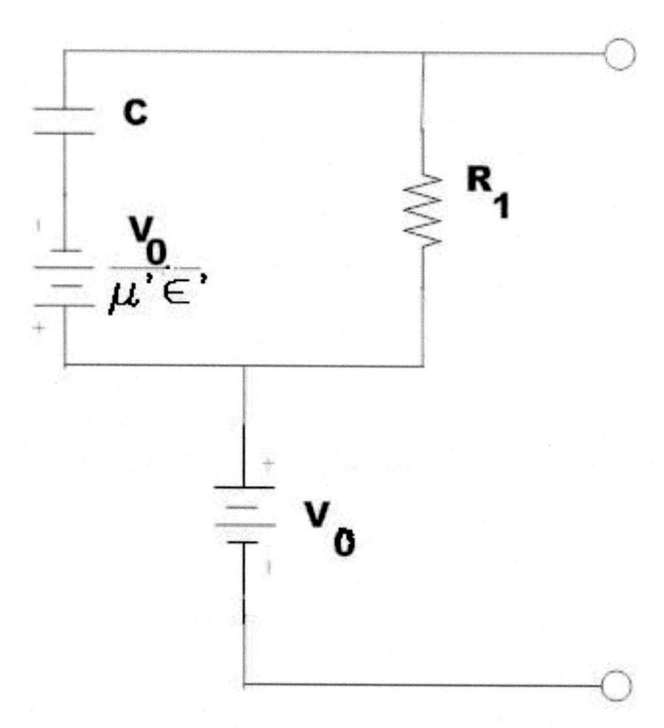

Figure 87: Equivalent circuit including losses

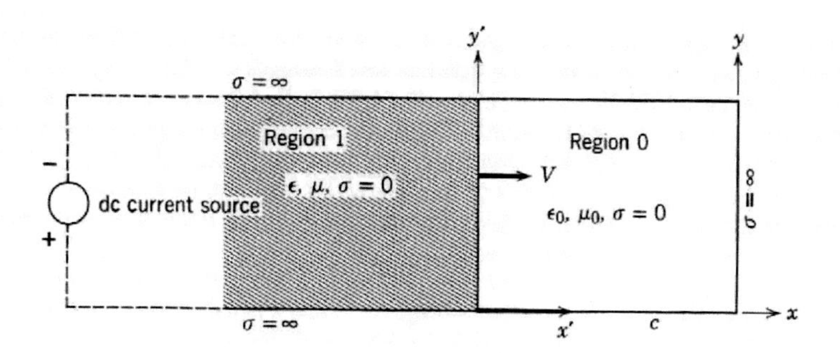

Figure 88: Conducting wall channel and moving dielectric

disc. The emf is given by the same general treatment of Eq.19 valid under the same restrictions as before regarding choice of path. It should be remembered that the terminal voltage must be the same regardless of the location or relative velocity of the observer. Show that if the metal disc is replaced by a disc of the same size made of a leaky, ferromagnetic dielectric with the inside and outside cylindrical surfaces copper plated, the equivalent circuit is given by Fig.87 where V_0 is the same as for the metal disc of Fig.86. C is the capacitance of the disc when considered a simple capacitor and $R_1$is the resistance between the inner and outer surfaces of the disc. The emf is given by the same general treatment of Eq.19 valid under the same restrictions as before regarding choice of path. It should be remembered that the terminal voltage must be the same regardless of the location or relative velocity of the observer.

This problem is of great theoretical importance because it seems to be a case in which the relativistic solution as given here and the classical solution differ even when the relative velocity is very small compared to that of light. The classical approach gives the battery in the lossy, magnetic dielectric case a value of V_0/ϵ' instead of $V_0/\mu'\epsilon'$.

10.5 Dielectric moving in a conducting walled channel

(a) A semi-infinite channel is made of very good conducting walls. It is shorted at $x = 0$ and is fed by a current source at $x = \infty$. The configuration is infinite in the x direction and is shown in Fig.88. A semi-infinite dielectric magnetic slab moves with velocity V in the **x** direction relative to the conducting walls of the channel. The dielectric magnetic slab is nonconducting and is characterized by permeability, μ and permittivity ϵ. Show that the fields observed in the laboratory frame,being the rest frame of the conducting plates, are given by

$$H_z^0 = I/[1 - (V^2/c^2)(\epsilon/\epsilon_0)]\beta^2$$

where I is the strength per unit width of the current source and $E_y^0 = D_y^0 = 0, B_z^0 = \mu_0 H_z^0$. In addition

$$E_y^1 = (\mu - \mu_0)\beta^2 V H_z^0$$

$$D_y^1 = -\mu_0(\epsilon - \epsilon_0)\beta^2 V H_z^0$$
$$B_z^1 = \mu[1 - (V^2/c^2)(\mu/\mu_0)]\beta^2 H_z^0$$
$$H_z^1 = [1 - (V^2/c^2)(\epsilon/\epsilon_0)]\beta^2 H_z^0$$

where the superscripts 1 and 0 refer to dielectric magnetic slab and free space respectively.

(b) Rework the problem for a finite (in the x dimension) dielectric magnetic slab.

(c) Rework (a) for the case of a conducting channel fed by a voltage source of strength V_b at $x = 0$ and open circuited at $x = \infty$. *Hint:*

$$H_z^0 = \epsilon_0 V_b V(1 - \epsilon/\epsilon_0)/\{d[1 - (V^2/c^2)(\epsilon/\epsilon_0)]\}$$

10.6 Mach's ten principles and general relativity

There may be more than ten but these are listed by Biondi and Samuel[10]. This is a review explaining the multiplicity of "Mach principles" which have been invoked in the research literature (and elsewhere). Ten are listed below:

Mach 0: The universe, as represented by the average motion of distant galaxies, does not appear to rotate relative to local inertial frames.

Mach 1: Newton's gravitational constant G is a dynamical field.

Mach 2: An isolated body in otherwise empty space has no inertia.

Mach 3: Local inertial frames are affected by the cosmic motion and distribution of matter.

Mach 4: The universe is spatially closed.

Mach 5: The total energy, angular and linear momentum of the universe are zero.

Mach 6: Inertial mass is affected by the global distribution of matter.

Mach 7: If you take away all matter, there is no more space.

Mach 8: is a definite number, of order unity, where is the mean density of matter in the universe, and is the Hubble time.

Mach 9: The theory contains no absolute elements.

Mach 10: Overall rigid rotations and translations of a system are unobservable.

10.6.1 Problems to solve

(a) The following question may shed some light on general relativity and Mach's principle. Consider two concentric spheres with equal and opposite charge constrained to remain uniformly distributed over their surfaces. When the spheres are at rest, there is no electric or magnetic field outside the outer sphere. When the spheres rotate with constant angular velocity about an axis through their center, there is again no electric field in the outer space region, but there is a magnetic field since the magnetic moment of each rotating sphere is proportional to the square of its radius. Calculate the value of this magnetic field. What are the fields seem by an observer in arbitrary relative motion to the rotating spheres? Suppose the spheres are stationary. Then, regardless of the motion of the observer, he should detect no field whatsoever. The covariance of Maxwell's equation implies that if both the electric and magnetic fields are zero in one frame of reference, they will remain zero in all reference frames. This is why, regardless of his frame of reference, the observer will find neither and electric nor a magnetic field since all the components of the electromagnetic field tensor are zero in one reference frame.

(b) According to Mach's principle, inertial effects are due to the distribution of mass in the universe. Hence, if the universe contained only two concentric spheres and an observer, there would be no way to find out whether or not the spheres were rotating. Einstein's general relativity does not give an unambiguous answer to the question of the possibility or impossibility of detecting the rotation motion of a single body in an otherwise empty universe. However, its answer tends toward the latter. A significant question is whether or not an observer could detect and electric or magnetic field from the concentric sphere system just described in an otherwise empty universe. The answer may seem obvious. If the spheres are at rest, regardless of the observer's motion he will detect no fields. There is no energy density in the surrounding space associated with either field, so how could the observer's motion affect the situation? If the spheres are rotating, the observer will detect either a magnetic or and electric field, or both, depending on the motion of his reference frame. Hence he could make an unambiguous conclusion regarding the rotation of the spheres. On the other hand, the question may not be so easy. The very reason why the observer can detect no field regardless of his motion when the spheres are stationary in the normal universe is apparently due to the presence of distant masses that might be charged. What happens when these masses are not present? Will it then be possible to detect rotation by this method?

10.7 Charge rearrangement when equatorial velocity approaches the speed of light

Investigate the distribution of total charge q, when the equatorial spin velocity v of a charged conducting sphere approaches c.

(a) Show that charge may be confined to a narrow equatorial belt at latitudes $b \leq \sqrt{3}(1 - v^2/c^2)^{1/2}$ while charge of the opposite sign occupies most of the sphere's surface. The change in field structure is shown to be a growing contribution of the "magic" electromagnetic field of the charged Kerr-Newman black hole with Newton's G set to zero.

(b) Show that the total charge, q_t within the narrow equatorial belt grows as $q_t \sim (1 - v^2/c^2)^{-(1/4)}$ and

$$\lim_{v \to c} (q_t) = \infty$$

(c) Calculate the electromagnetic field, Poynting vector, field angular momentum, and field energy.

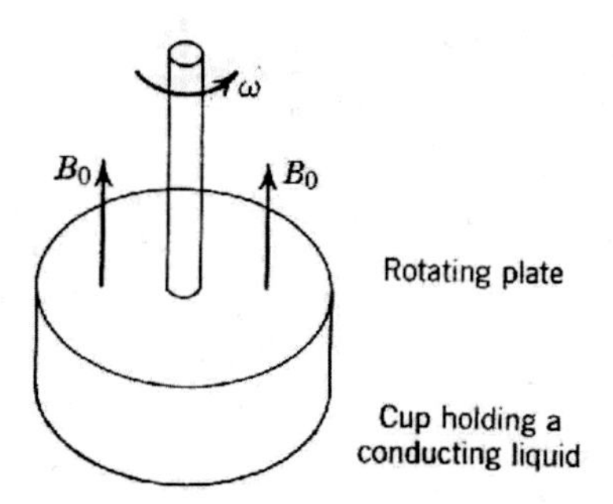

Figure 89: A circular MHD Coutte device

See papers by Newman and Janis[11] and Newman, Chinnapared, Exton, Prakash and Torrence[12].

10.8 A circular MHD Couette device

Fig.89 shows a circular MHD Couette device similar to the Faraday disc. A shaft rotates the top insulating plate with an angular velocity of ω. The bottom portion of the device comprises a cup holding a conducting liquid. The bottom of the cup is an electrical insulator. The circular wall of the cup may be an ideal conductor or insulator. The top insulating plate drags the conducting fluid through the uniform axial magnetic field, B_0. Find the velocity profile, the current density vectors and the emf induced between the axis of the shaft and the cup wall for an (a) ideally insulating cup wall and (b) ideally conducting cup wall.

Part VII
Appendices

A Distortion of magnetic induction field as portrayed in Example 6.7

In this and subsequent appendices the finite element software of www.PDEsolutions.com is used. Student and trial professional versions of the software is available at the aforementioned web site. There are many other finite element programs available which are so specialized that only certain kinds of preset problems can be solved. Many of the applications to electrostatics are discussed in a published book[13]. Here the two dimensional problem showing how the presence of a highly conducting disc can change the distribution of magnetic induction or flux density. The highly conducting disc is placed between two cylinders of moderate conductivity. Magnetic flux is forced to flow axially along the cylinders. When it encounters the highly conducting disk it becomes concentrated in the outer layers of the disk. This problem is solved using the finite element software called FlexPDE. A demo and a free student version can be obtained on the Internet at the URL www.pdesolutions.com. Below is the input file with explanation following the ! glyphs.

```
TITLE 'Distortion in magnetic induction by conducting Disc'
COORDINATES
xcylinder(z,r) !Request for cylindrical coordinates
VARIABLES !Variables defined
Br  Bi
SELECT
errlim= 1e-05 !Setting error limit
DEFINITIONS ! Defining parameters
delta !Skin depth to be set in various regions
gap = 0.2 !Air gap
EQUATIONS
Br: curl(curl(Br))-(2/delta^2)*Bi = 0
Bi: curl(curl(Bi))+(2/delta^2)*Br = 0
BOUNDARIES
Region 1 delta=1000 !Low conductivity
start(0,0) line to (0,25) to (25,25) to (25,0) to finish
Region 2 delta=100 !Moderate conductivity
start(2,0) line to ( 2,2) value(Br)=1 value(Bi)=0 line to (10,2)
natural(Br)=0 Natural(Bi)=0 line to (10,0) to finish
Region 3 delta=0.3 !High conductivity
start(0,0) line to (0,2) value(Br)=1 value(Bi)=0 line to (2-gap,2)
natural(Br)=0 Natural(Bi)=0 line to (2-gap,0) to finish
MONITORS
vector(Br) vector(Bi)
PLOTS !Requesting plots of magnetic induction
vector(Br) zoom(0,0,4.2,4.2) as 'Re magnetic induction'
vector(Bi) zoom(0,0,4.2,4.2) as 'Im magnetic induction'
contour((Br^2+Bi^2)^0.5) zoom(0,0,4.2,4.2) as
'magnitude of magnetic induction'
END
```

The magnetic induction is given by Maxwell's equations. Let the magnetic induction be given by

$$\mathbf{B}(x, y, z, t) = \mathbf{B}(\mathbf{x}, \mathbf{y}, \mathbf{z})e^{j\omega t} \tag{191}$$

Relationships like this are used to transform Maxwell's equations to make it easy to solve problems involving sinusoidal excitation. Maxwell's equations become

$$\nabla \times \mathbf{E} = -j\omega\mathbf{B} \tag{192}$$

and

$$\nabla \times \mathbf{B} = \sigma\mu\mathbf{E} \tag{193}$$

noting that the permeability has been removed from the curl operation because it is constant everywhere in this problem. Taking the curl of Eq.193 and using Eq.192 there results

$$\nabla \times \nabla \times \mathbf{B} + j\omega\sigma\mu\mathbf{B} = \mathbf{0} \tag{194}$$

Because FlexPDE doesn't not accept complex quantities as variables this equation must be written in terms of its real and imaginary parts. The results are given by

$$\nabla \times \nabla \times \mathbf{B}_r - (2/\delta^2)\mathbf{B}_i = 0 \tag{195}$$

and

$$\nabla \times \nabla \times \mathbf{B}i + (2/\delta^2)\mathbf{B}_r = 0 \tag{196}$$

where $\delta = (\omega\sigma\mu/2)^{-1/2}$. It is interesting to note that in well annealed pure copper at 60 Hz the skin depth is given by $\delta = (100/2.54) \times (2\pi \times 60 \times 5.8 \times 10^7 \times 4\pi \times 10^{-7}/2)^{-1/2} = 0.335\,89$ inches. This is why very few solid thick copper wires are used commonly. These equations are written in the section under 'EQUATIONS' and the geometry is set in the section entitled 'BOUNDARIES'. Included are the necessary boundary conditions. Several plots are also specified showing the vectors of $\mathbf{B}_r$ and $\mathbf{B}_i$ .It is interesting to note how the real and imaginary parts of the magnetic induction interact to alter the magnitude of the magnetic induction in the conductors.

B Faraday disc in a nonuniform magnetic field as portrayed in Example 6.8

Here the two dimensional problem showing the nonuniform magnetic field passing through a Faraday disc is presented. This is a cylindrical geometry containing a permanent ring magnet in the cylinder located at the right. Only the first quadrant of the geometry needs to be considered in cylindrical coordinates. There is a top plate and inner cylinder to complete the path taken by the magnetic flux through the air gap in the lower left part of the geometry. The rotating Faraday disc would be located in the air gap and it is assumed to not be ferromagnetic. A slight modification can be made by the ambitious reader to make it ferromagnetic with a small air gap located between it and the magnetic pole. The geometry is given in Fig.90. This is solved by using the following equations:

$$\nabla \times \mathbf{H} = 0 \tag{197}$$

$$\mathbf{H} = [\nabla(\mathbf{A}) - \mathbf{P}]/\mu \tag{198}$$

and

$$\mathbf{P} = \mathbf{z}P_z$$

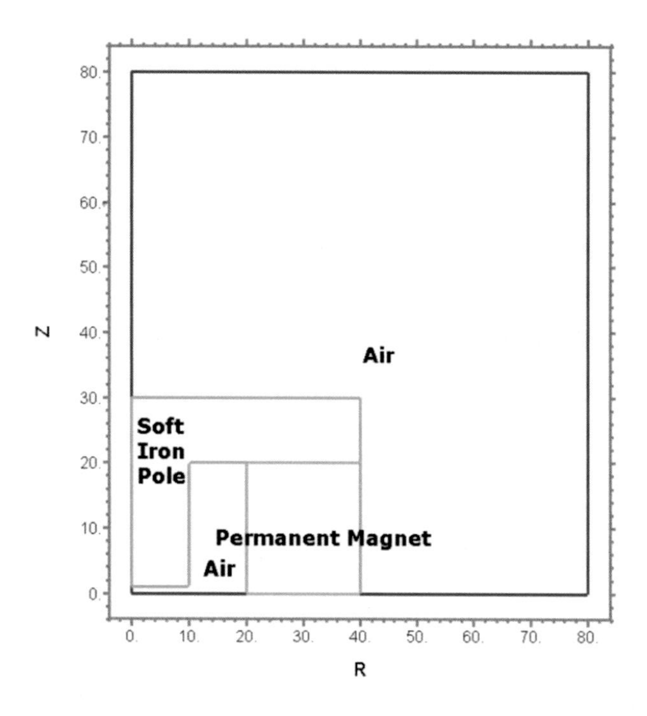

Figure 90: Faraday disc magnet geometry

and $\mathbf{A} = A_z$. A Neuman boundary condition on **H** is used on the outer boundaries to invoke symmetry on R=0 and Z=0. The other outer boundaries are chosen to be far enough away from the magnet structure that a Neuman boundary condition which is not correct will not harm the results in the vicinity near the origin. These equations are solved simultaneously to produce the various plots called in the PLOT section of the script.

```
{ PERMANENT_MAGNET1.PDE }
Title 'PERMANENT-MAGNET'
Select
errlim = 1e-6
coordinates
ycylinder(r,z)
Variables
A      { z-component of Vector Magnetic Potential }
Definitions
mu
Pr = 0 { Magnetization components }
Pz = 0
```

```
P = vector(Pr,Pz) { Magnetization vector }
H = (curl(A)-P)/mu { Magnetic field }
z0 = 1 { Air Gap Thickness }
Initial values
A = 0
Equations
curl(H) = 0
Boundaries
Region 1
mu = 1
start(0,0) line to (80,0)
value(A)=0 line to (80,80) to (0,80) to finish
Region 2
mu = 10000
start(0,z0)line
to (10,z0) fillet(0.5) to (10,20) fillet(0.5)
to (40,20) to (40,30) to (0,30) to finish
Region 3 { the permanent magnet }
mu = 1
Pz = 100
start (20,0) line to (40,0) to (40,20) to (20,20) to finish
Monitors
Contour(A)
Plots
grid(r,z) zoom(0,0,40,30) as 'Grid for the iron parts and air gap' contour(A)
painted zoom(0,0,42,42)
vector(curl(A)) zoom(0,0,42,42) as 'FLUX DENSITY B'
elevation(ycomp(curl(A))) from (1,0) to (16,0) as 'Bz'
elevation(r*ycomp(curl(A))) from (1,0) to (16,0) as 'r*Bz'
Table(r*ycomp(curl(A)),r) from (1,0) to (16,0)
vector(H) zoom(0,0,42,42) as 'MAGNETIC FIELD H'
vector(H) zoom(25,0,10,10) as 'MAGNETIC FIELD H'
contour(A) zoom(0,0,42,42) as 'Az MAGNETIC POTENTIAL'
vector(P) zoom(25,0,10,10) as 'MAGNETIC POLARIZATION'
End
```

C Magnetic induction in the vicinity of an air hole in Example 6.11

Here the magnetic vector potential is solve in the first quadrant of the z = 0 plane and the adjoint partial differential equations are solved to obtain equipotential lines. This program calculated the distribution of the magnetic induction in iron and circular holes which might hold conductors. Although the conductors could have relative permeability greater than unity the permeability of air is used in the conductor holes. For example, copper is paramagnetic and has a relative permeability as large as 1.06. If ferrous conductors were used, their permeability could be added easily. The purpose of this calculation is to show how the magnetic induction in air holes diminishes as the permeability of the surrounding iron increases. This is done by calling for 9 different permeabilities in the second line of SELECT. In the second line of **DEFINITIONS** the relative permeability is defined. Note that more than one parameter can be defined on any given line. The next four lines calculate data for hole locations. Here the vector potential is defined as V. The boundary conditions have been chosen so that given no holes the magnetic induction would be radial and appropriate symmetry would yield the first quadrant of the z = 0 plane. Features are defined so that magnetic induction can be plotted along circumferences drawn through the center of the conductor holes and along the iron-air boundary. Many interesting plots are made which can be shown in Flexpde as movies. Unfortunately they cannot be included here. However, a few are given as figures. Some of the most instructive are the small curvilinear rectangle plots for various values of relative permeability. Also exhibited here are plots of the magnetic induction vectors in the vicinity of a conductor hole for various k.

TITLE 'Flux in Conductor Holes'
COORDINATES cartesian2
VARIABLES
V Q ! V is the magnetic vector potential and Q the adjoint vector potential
SELECT
errlim = 1e-04
stages=9
DEFINITIONS
Mu0 = Pi*4E-7
Mu k = staged(2.5,5,7.5,10,12.5,15,17.5,20,25)
x1 = 0.75*cos(Pi/6) x2 = 0.75*cos(Pi/3)
x3 = 0.80*cos(Pi/6) x4 = 0.80*cos(Pi/3)
y1 = 0.75*sin(Pi/6) y2 = 0.75*sin(Pi/3)
y3 = 0.80*sin(Pi/6) y4 = 0.80*sin(Pi/3)
B = curl(V)
EQUATIONS
V:curl(curl(V)/Mu) = 0
Q:Div(Grad(Q)*Mu) = 0 ! adjoint equation to calculate magnetic equipotentials
BOUNDARIES
Region 'Outer Air' Mu = mu0
start(0,1.2) value(V) = 1 natural(Q) = tangential(grad(V)) line to (0,1)
line to (0,0.2) natural(V) = 0 arc(center= 0,0) angle = -90 value(V) = 0 line to (1.2,0)
natural(V) = 0 arc(center = 0,0) angle = 90 to finish
Region 'Iron' Mu = k*mu0
start(0.2,0) value(V) = 0 natural(Q) = tangential(grad(V)) line to (0.75,0)
natural(V) = 0 arc(center=0.8,0) angle=-180 line to (1,0) arc(center=0,0)

```
angle = 90 value(V) = 1 line to (0,0.85) natural(V) = 0 arc(center=0,0.8)
angle = -180 value(V) = 1 line to (0,0.2) natural(V) = 0 arc(center=0,0) to
finish
feature 'circle 1'
start(0.8,0) arc(center = 0,0) angle = 90
feature 'circle 2'
start(1,0) arc(center = 0,0) angle = 90
Region 'air Hole1' Mu = Mu0
start(x1,y1) natural(Q) = tangential(grad(V)) arc(center = x3,y3) angle =
360
Region 'air Hole2' Mu = Mu0
start(x2,y2) natural(Q) = tangential(grad(V)) arc(center = x4,y4) angle =
360
Region 'outer iron' mu = k*mu0
start(1.02,0) natural(Q) = tangential(grad(V)) line to (1.1,0)
arc(center=0,0) angle = 90 line to (0,1.1) to (0,1.02)
arc(center=0,0) angle = -90 to finish
MONITORS
contour(V,Q)
PLOTS
grid(x,y) zoom(-0.02,-0.02,1.22,1.22) as 'Finite element mesh'
contour(V,Q) zoom(-0.02,-0.02,1.22,1.22) vector(B) zoom(-0.02,-0.02,1.22,1.22)
vector(B) zoom(0.6,0.3,0.2,0.2) as 'Magnetic induction vectors'
elevation(ln(((x^2+y^2)^0.5)*magnitude(B))*cos(arctan(ycomp(B)/xcomp(B))-
arctan(y/x)))
on 'circle 2' as 'Ln(B radial) on rotor outer circle'
elevation(ln(((x^2+y^2)^0.5)*magnitude(B))*cos(arctan(ycomp(B)/xcomp(B))-
arctan(y/x)))
on 'circle 1' as 'Ln(B radial) on hole center circle'
contour(Mu) painted
END
```

Figs.91 , 92 and 93 show the flux tubes avoid the holes more as k increases. Figs.94, 95 and 96 more clearly show the induction vectors in the holes diminishing as k increases. In iron used in most motors and generators has permeability such that $1000 \leq k \leq 1000000$ which means there may be almost no magnetic induction in the conductor holes.

D Two permanent magnet generators considered in Ex. 6.12

Here is the script for the two two-dimensional permanent magnet generators portrayed in the x-y plane and very long in the z dimension. The one on the left has a smooth rotor whilst the other has a square conductor hole in its rotor. Both are excited by a magnet located in the upper center. This program solves

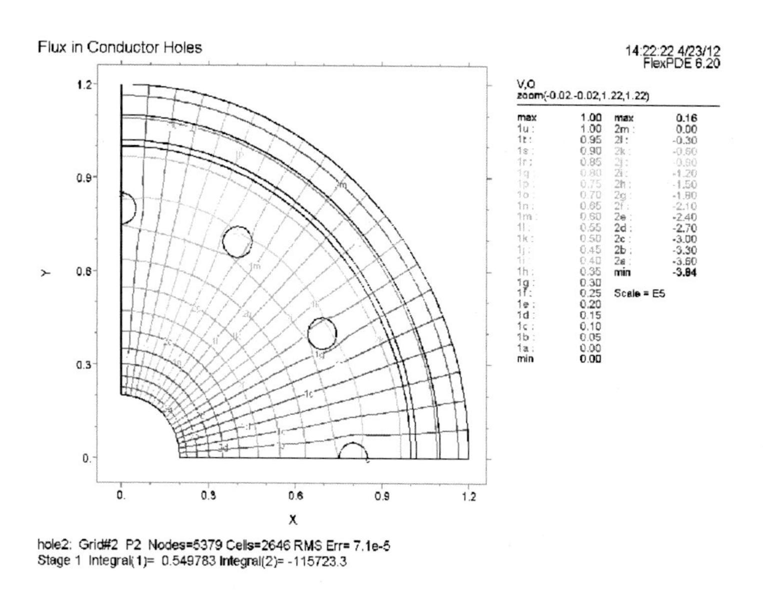

Figure 91: Curvelinear plot for k = 2.5

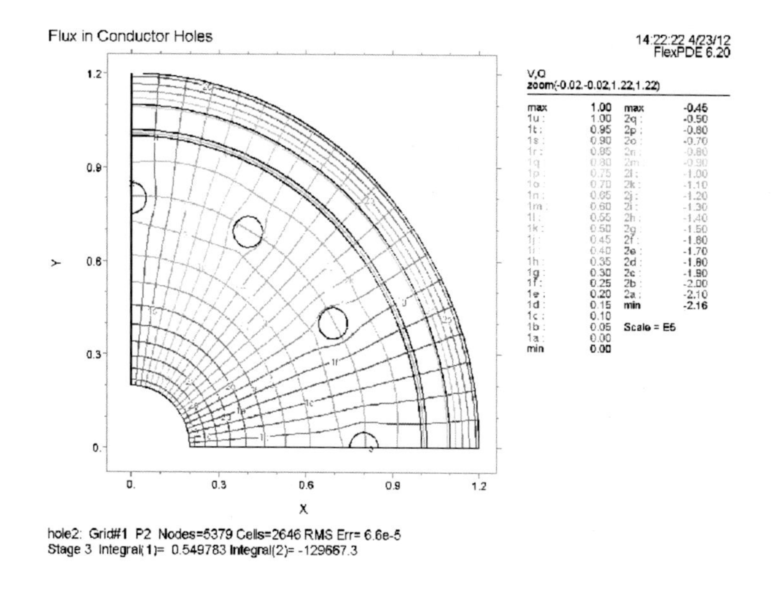

Figure 92: Curvelinear plot for k = 7.5

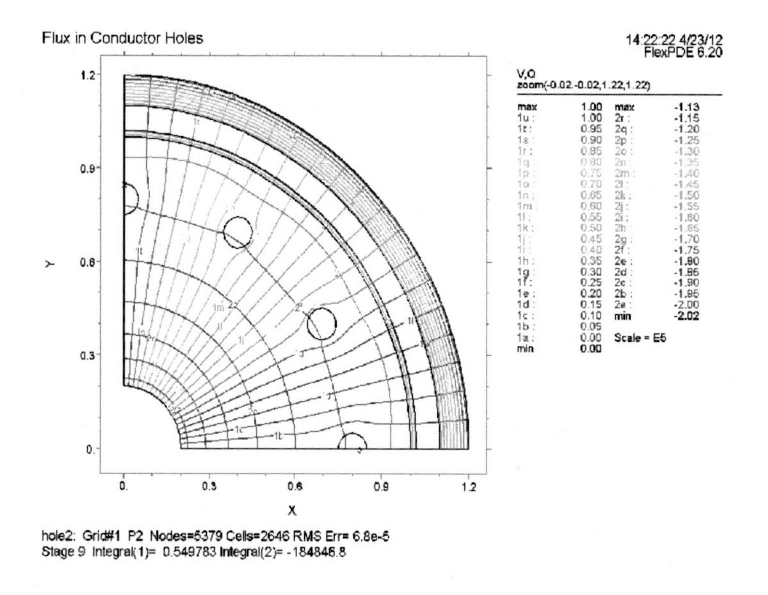

Figure 93: Curvelinear plot for k = 20

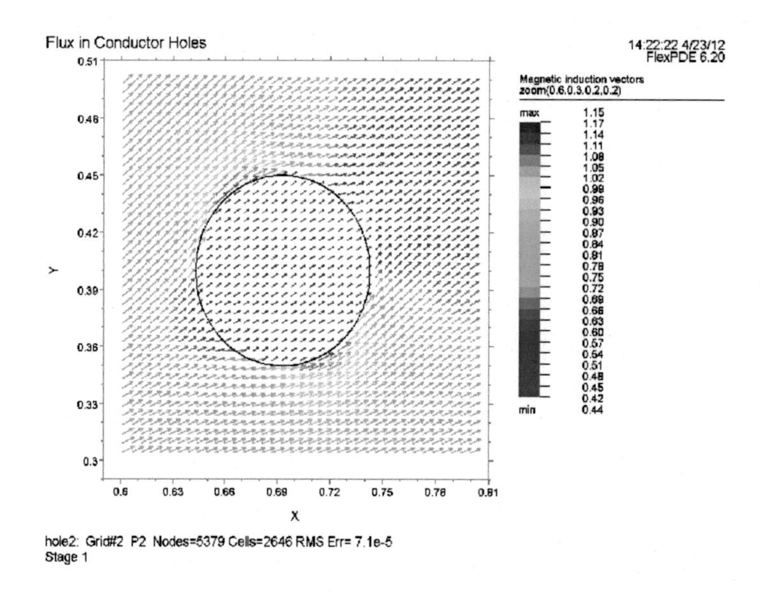

Figure 94: Magnetic induction vectors for k = 2.5

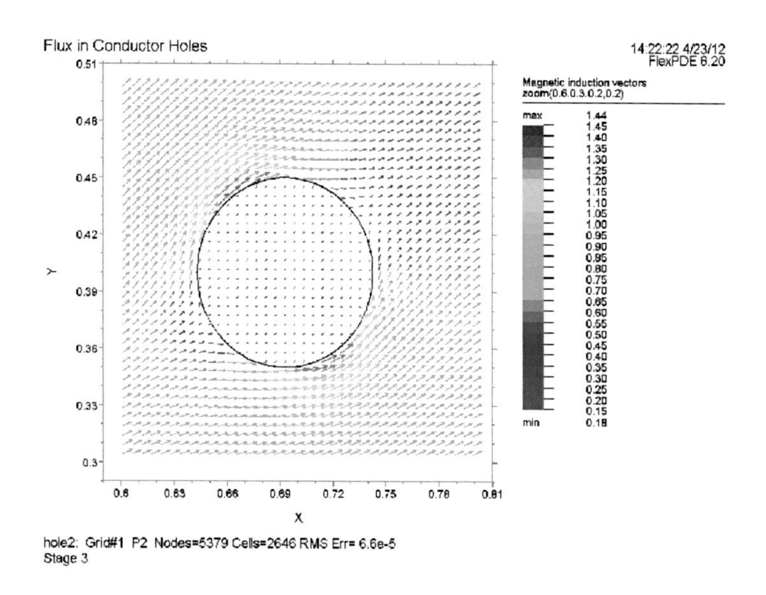

Figure 95: Magnetic induction vectors for k = 7.5

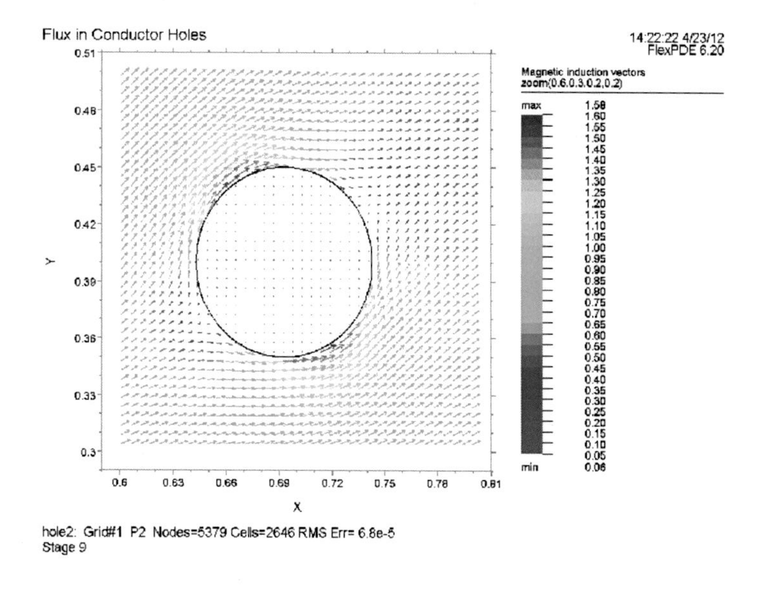

Figure 96: Magnetic induction vectors for k = 20

for the vector potential, A_z , **B** and **H** as in the earlier scripts. Some of the results are given in Example 6.12.

```
{ PERMANENT_MAGNET_DC_GENERATOR.PDE }
Title 'A PERMANENT-MAGNET DC Generator'
Select
errlim = 1e-6
ymergedist = 0.00254 ymergedist = 0.00254
featureplot = on
Coordinates
cartesian2
Variables
A ! z-component of Vector Magnetic Potential
Definitions
mu m = 0.0254 mu0 = Pi*4E-07
Gp = 0.15
Mx = 0 { Magnetization components }
My = 0
MM = vector(Mx,My) { Magnetization vector }
H = ((curl(A)/mu)-MM) { Magnetic field }
B = curl(A)
Initial values
A = 0
Equations
curl(H) = 0
Boundaries
Region 1 'Outer Air' mu = mu0
start(-20*m,0) line to (20*m,0) value(A) = 1 line to (20*m,40*m) to (-
20*m,40*m) to finish
Region 2 'Stator' mu = 10000*mu0
start(20*m,(10+Gp)*m) line to (20*m,40*m) to (-20*m,40*m) to (-20*m,(10+Gp)*m)
arc(center=-20*m,0) angle = -60 line to (-0.5*m*(10+Gp),15*m) to (0.5*m*(10+Gp),15*m)
to ((20-0.866*(10+Gp))*m,0.5*m*(10+Gp)) arc(center=20*m,0) to finish
Region 3 'Permanent Magnet' mu = mu0 Mx = 1000
start(-2*m,40*m) line to (-2*m,15*m) to (2*m,15*m) to (2*m,40*m) to fin-
ish
Region 4 'Left Rotor' mu = 10000*mu0
start(-20*m,10*m) line to (-20*m,0) to (-10*m,0) arc(center=-20*m,0) angle
90
Region 5 'Right Rotor' mu = 10000*mu0
start(10*m,0) line to (20*m,0) to (20*m,10*m) arc(center=20*m,0) angle 90
Region 6 'Slot in Right Rotor' mu = mu0
start((20-4.5114)*m, 8.9245*m) line to ( (20-4.0603)*m,8.0321*m) arc(center=20*m,0)
angle = 6.3661
line to ((20-5.4732)*m, 8.3693*m) arc(center=20*m,0) angle = -6.3661
Feature 'Right Rotor Surface'
start(20*m,10*m) arc(center = 20*m,0) angle = 60
```

Feature 'Left Rotor Surface'
start(-20*m,10*m) arc(center = -20*m,0) angle = -60
Feature '1/2 under Right Rotor Surface'
start(20*m,9.5*m) arc(center = 20*m,0) angle = 60
Feature '1/2 under Left Rotor Surface'
start(-20*m,9.5*m) arc(center = -20*m,0) angle = -60
Monitors
contour(A)
Plots
contour(A) as 'Flux Tubes in the Upper Quarters of Two DC Generators'
contour(A) zoom(-20*m,0,20*m,20*m) as 'Flux Tubes in Left Side'
contour(A) zoom(0,0,20*m,20*m) as 'Flux Tubes in Right Side'
vector(H) zoom(-20*m,0,20*m,20*m) as 'Magnetic Intensity on Left Side'
vector(B) zoom(-20*m,4*m,10*m,10*m) as 'Magnetic Induction Vectors in Left Rotor'
vector(B) zoom(10*m,4*m,10*m,10*m) as 'Magnetic Induction Vectors in Right Rotor'
vector(B) zoom(0.36,0.18,0.06,0.06) as 'Magnetic Induction Vectors in Right Rotor Slot'
vector(B) zoom(-0.42,0.18,0.06,0.06) as 'Magnetic Induction Vectors in Left Rotor'
elevation(normal(B)) on 'Right Rotor Surface'
elevation(normal(B)) on 'Left Rotor Surface'
elevation(normal(B)) on '1/2 under Right Rotor Surface'
elevation(normal(B)) on '1/2 under Left Rotor Surface'
End

E Glass plate of Example 10.3

This is the flexpde script for solving the fields in and around the of example 10.3. Here it is convenient to obtain the electric field intensity, $\mathbf{E}$ from $\nabla \times \mathbf{zA}_z$. This is valid because the magnetic induction $\mathbf{B}$ does not vary in time.

{Glass Plate Moving in a Transverse Magnetic Intensity Field.}
TITLE 'Glass Plate Moving in a Transverse Magnetic Field'
COORDINATES cartesian2
VARIABLES
A ! z component of vector potential
SELECT
errlim=1e-06
DEFINITIONS
ep0 = 8.8541853E-12
V = 1 !meter/second
B0 = 1 !10 kilogauss
k = 5 Px = 0 Py = 0
P = vector(Px,Py)

```
d = 0.1
eps
E = (curl(A)/eps-P) Ex = xcomp(E) Ey = ycomp(E)
DD = eps*(E - P) DDx = xcomp(DD) DDy = ycomp(DD)
EQUATIONS
curl(E) = 0
BOUNDARIES
Region 1 eps=ep0
start(-20,-5) value(A)=0 line to (-5,-5) TO (5,-5) to (20,-5) to (20,5) to (-
20,5) to close
REGION 2 eps = k*ep0 Py = B0*V
START(-0.5,d/2) LINE TO (-0.5,-d/2) line to (0.5,-d/2) TO (0.5,d/2) TO
CLOSE
region 3 eps = 1.5*ep0
start(-d/20,-d/2) line to (-d/20,-3*d/2) to (-0.6,-3*d/2) to (-0.6,3*d/2) to
(-d/20,3*d/2) to (-d/20,d/2)
to (d/20,d/2) to (d/20,8*d/5) to (-0.6-d/10,8*d/5) to (-0.6-d/10,-8*d/5) to
(d/20,-8*d/5) to (d/20,-d/2)
to close
MONITORS
contour(A) zoom(-0.65,-2*d,0.65+d/5,4*d)
vector(E) zoom(-0.65,-2*d,0.65+d/5,4*d)
contour(eps/ep0) zoom(-0.65,-0.65,1.3,1.3) painted
PLOTS
contour(eps/ep0) zoom(-0.65,-0.65,1.3,1.3) painted
vector(E) zoom(-0.65,-0.65,1.3,1.3)
vector(DD) zoom(-0.65,-2*d,0.65+d/5,4*d)
contour(A) zoom(-0.65,-0.65,1.3,1.3)
vector(E) zoom(-0.65,0,0.65+d/5,1.8*d)
vector(E) zoom(-0.65,0,0.14,0.1)
vector(E) zoom(-d,-d,2*d,2*d)
elevation(Ey) from (-0.6005,d) to (-0.6005,-d)
elevation(Ey) from (0,-d/1.99) to (0,d/1.99)
elevation(Ey) from (-0.5,0) to (0.5,0)
elevation(DDy) from (-0.65,0) to (0.65,0)
elevation(DDy) from (0,1.2*d) to (0,-1.2*d)
END
```

F Three dimensional solution to Foucault's and Le Roux's apparatus

This author could not find the dimensions of Faraday's original or Foucault's and Le Roux's generator. Instead they have been assumed to be more or less in proportion to those in Fig. 9. Before describing the three dimensional analysis, it is necessary to change the form of the partial differential equations necessary to solve these problems using flexPDE software.

F.1 The divergence form[14,15,16]

In two dimensional geometry with a single nonzero component of $\mathbf{A}$, the guage condition $\nabla \cdot \mathbf{A} = 0$ is automatically satisfied and $\nabla \times \mathbf{H} = \mathbf{J}$ in the absence of displacement currents and and time varying currents. Letting $\mathbf{B} = \nabla \times \mathbf{A}$ yields the standard equation for solving magnetic problems, and is given by

$$\nabla \times [(\nabla \times \mathbf{A})/\mu] = \mathbf{J} \tag{199}$$

In cases of multiple materials it is mandatory to keep the permeability, μ inside the curl operator to give the proper boundary conditions at the interfaces between different materials, the result being $\mathbf{n} \times \mathbf{H}$ is continuous across internal boundaries. However, in three dimensions Eq.199 does not impose a gauge condition , and is ill posed in many cases. There exists a gauge condition such that

$$\nabla \times [(\nabla \times \mathbf{A})/\mu] = -\nabla \cdot [(\nabla \mathbf{A})/\mu] \tag{200}$$

It can be shown that Eq.199 becomes

$$\nabla \cdot [(\nabla \mathbf{A})/\mu] + \nabla \times \mathbf{M} + \mathbf{J} = 0 \tag{201}$$

F.2 The boundary conditions

The conversion to the divergence form modifies the interface conditions. The Neuman boundary condition for each component of Eq.201 is now the normal component of the argument of the divergence such that

$$\begin{aligned} Natural(A_x) &= \mathbf{n} \cdot (\nabla A_x/\mu) \\ Natural(A_y) &= \mathbf{n} \cdot (\nabla A_y/\mu) \\ Natural(A_z) &= \mathbf{n} \cdot (\nabla A_z/\mu) \end{aligned} \tag{202}$$

FlexPDE does not integrate constant source terms by parts, and if $\mathbf{M}$ is piecewise constant the magnetization term will disappear in the equation analysis. Thus it is necessary to reformulate the magnetic term so that it can be incorporated into the divergence. It follows that

$$\iiint_V \nabla \times \mathbf{M} dV = \iint_S \mathbf{n} \times \mathbf{M} dS \tag{203}$$

and

$$\mathbf{n} \times \mathbf{M} = \mathbf{n} \cdot \mathbf{N} \tag{204}$$

which is formed by defining $\mathbf{N}$ as the antisymmetric dyadic

$$\mathbf{N} = \begin{pmatrix} 0 & M_z & -M_y \\ -M_z & 0 & M_x \\ M_y & -M_x & 0 \end{pmatrix} \tag{205}$$

then because integration by parts will produce surface terms $\mathbf{n} \cdot \mathbf{N}$, that are equivalent to the required surface terms $\nabla \times \mathbf{M}$.

F.3 The equations to solve

Due to the aforementioned requirements the equations to be solved are rather different looking than those used in two dimensional problems. They are equivalent and must be used because of the properties of the finite element solver. Then Eq.201 becomes

$$\nabla \cdot \{[(\nabla \mathbf{A})/\mu] + \mathbf{N}\} + \mathbf{J} = 0 \tag{206}$$

Because flexPDE does three dimensional problems in Cartesian coordinated Eq.206 is written in as the following three equations:

$$\begin{aligned} \nabla \cdot \{[(\nabla A_x)/\mu] + N_x\} + J_x &= 0 \\ \nabla \cdot \{[(\nabla A_y)/\mu] + N_y\} + J_y &= 0 \\ \nabla \cdot \{[(\nabla A_z)/\mu] + N_z\} + J_z &= 0 \end{aligned} \tag{207}$$

where $\mathbf{N}_i$are the rows of $\mathbf{N}$. Here the Neuman boundary condition will be defined as the value of the normal component of the divergence such that

$$Natural(A_x) = \mathbf{n} \cdot [(\nabla A_x/\mu) + N_x] \tag{208}$$

In this problem there is only M_z applied to excite the Faraday generator resulting in $A_z = 0$. In this case solutions for the x and y components of $\mathbf{A}$ are sought in the EQUATIONS section of this script.

F.4 Geometry

It is more difficult to set the geometry in three than in two dimensions. Flexpde accomplishes this by extruding two dimensional shapes. It is rather like stacking blocks or coins to make three dimensional objects from a collection of two dimensional shapes. The approximate geometry of the Foucault and Le Roux generator is exhibited in Fig.97. By making use of symmetry only the lower half of the generator need be considered. At the top is the iron disc that rotates. The magnetic field is provided by a permanent magnet and a pole piece structure to make magnetic flux pass through the disc in approximately the z direction. The red line on the iron disc running from the center to the left edge of the disc is a feature needed to allow the z component of the magnetic induction to be plotted. This is significant because most early investigators assumed

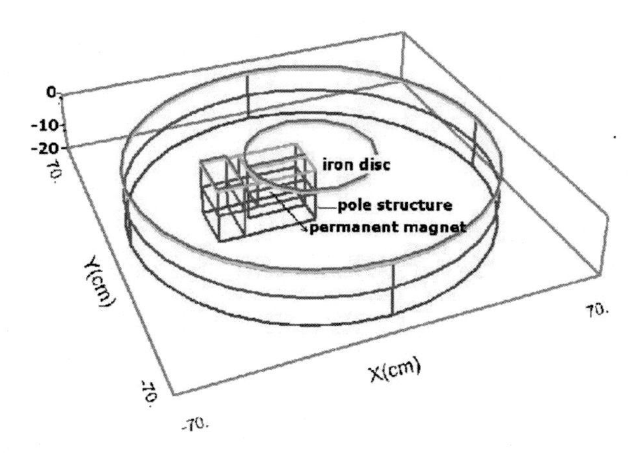

Figure 97: Lower half of the generator

it constant. Here that assumption is checked. The extrusion must always start from the bottom of the geometry. The bottom 10 cm is air followed by 8.9 cm of iron space. Next is 1 mm of air followed by 1 cm of iron for the disc. These layers must be properly cited in the boundaries section of the script. The next two figures show the log to the base ten of the permeability of the various parts of the geometry. Here it must be remembered that the permeability of the permanent magnet has been chosen to be that of air in light of the lack of knowledge of the kind of material used by Foucault and Le Roux. A block just below the air gap below the disc is the permanent magnet. Here it is assumed the shaft is nonmagnetic although if it were of copper it might be paramagnetic with $\mu \simeq 1.06\mu_0$. In Fig.98 The permanent magnet is located in the air above the x and fills that region. The region where the o is located is air between the permanent magnet and the iron pole piece. Another view is shown in Fig.99.

In Fig.100 the magnetic induction vectors are shown on the plane $y = 0$. The third quadrant of that plane is shown for the variables $-26 \leq x \leq 0$ & $-17 \leq z \leq 0$. In this depiction B_z looks almost constant in the disc and B_x

72

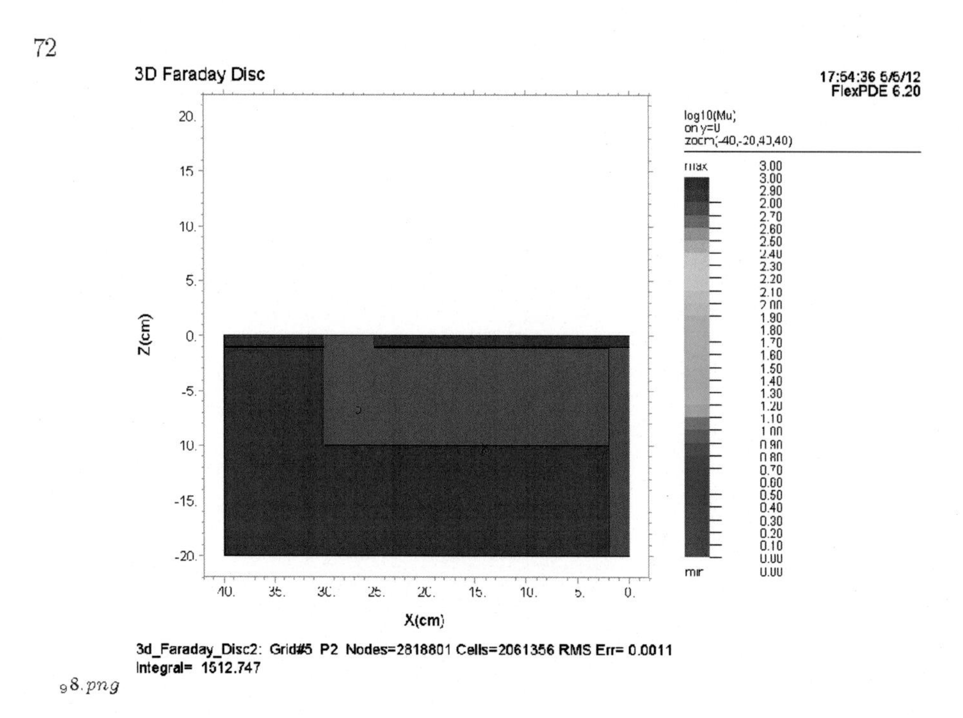

98.png

Figure 98: The geometry on the plane $y = 0$

73

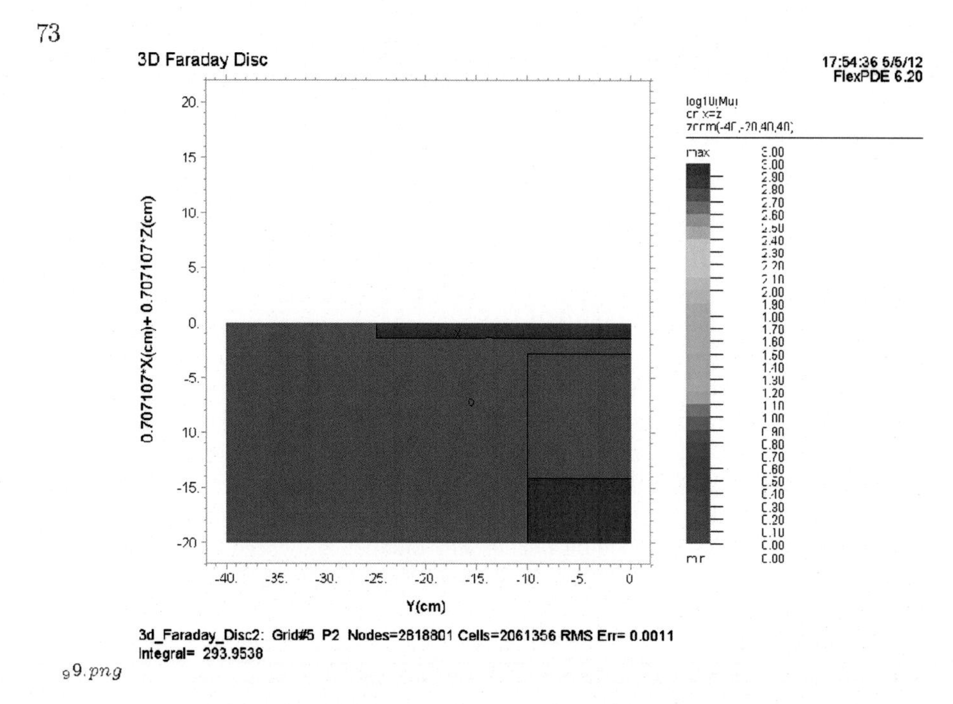

99.png

Figure 99: The geometry on the plane $x = z$

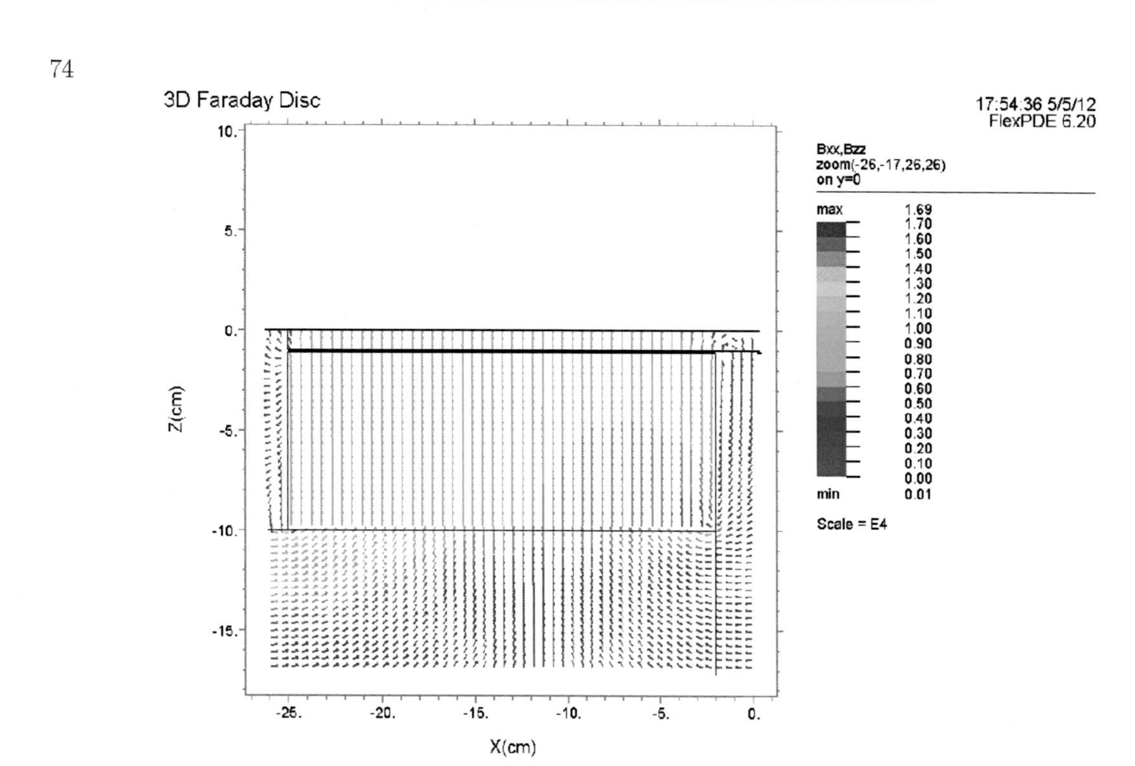

100.png

Figure 100: Magnetic induction vectors in disc

seems much smaller. Fig.101 show the contours of B_z on the $z = 0$ plane which passes through the center of the disc in the radial direction. The said magnetic induction is almost constant over much of the area above the permanent magnet. In Fig.102 the actual variation of B_z is exhibited on the $z = 0$ plane for $(-25, 0) \leq (-2, 0)$. If one wanted to find the induced emf it would have been easy to change the last instruction in the script to *elevation(Ω ∗ x∗Bzz)*. In that case the integral appearing at the lower left corner of the figure would be $\int_{-25}^{-2} \mathbf{V} \times \mathbf{B} \cdot dl$. In conclusion, this solution may tax many present day personnel computers because to get an RMS error of 0.0011 required a mesh containing 2,061,356 cells and 2,818,801 nodes. The run time was several hours.

G The FlexPDE script for a 3D solution to the Faraday disc

TITLE '3D Faraday Disc '

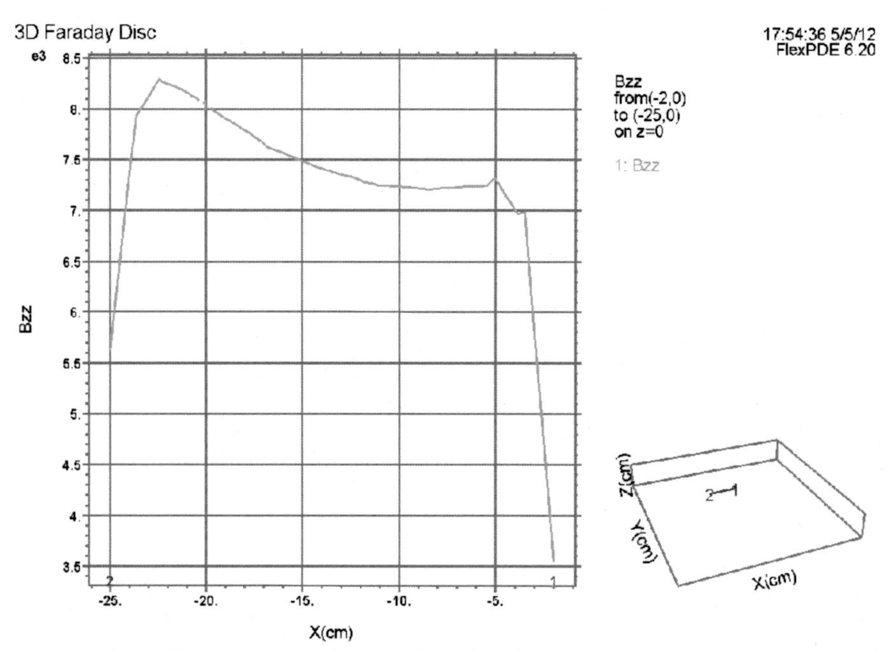

Figure 101: B_z on the $z = 0$ plane

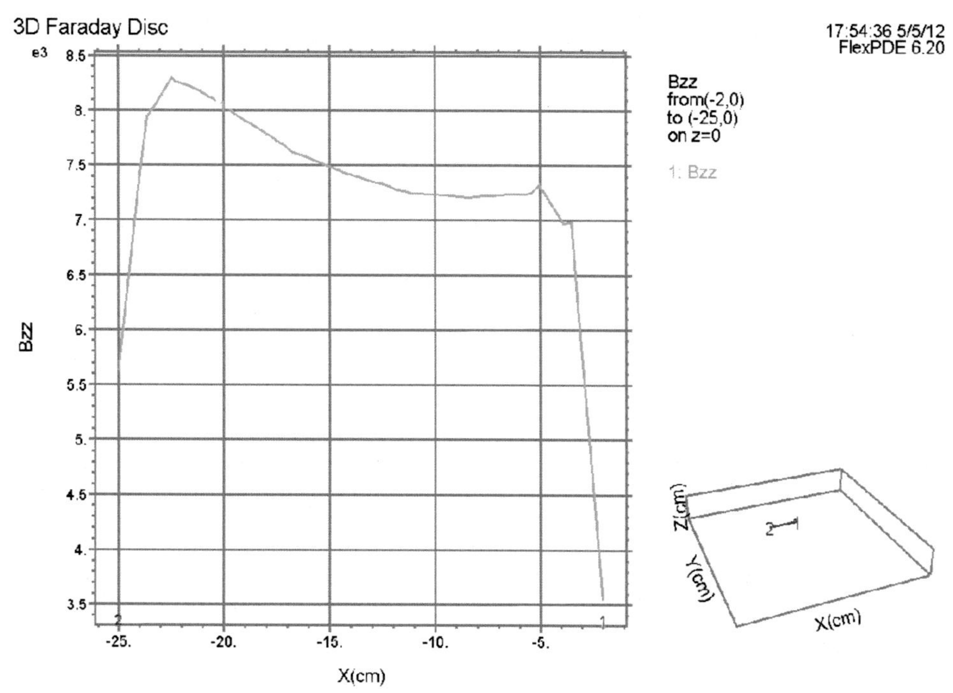

Figure 102: B_z on the $z = 0$ plane

```
COORDINATES
CARTESIAN3
SELECT
alias(x) = "X(cm)"
alias(y) = "Y(cm)"
alias(z) = "Z(cm)"
VARIABLES
Ax,Ay { assume Az is zero! }
DEFINITIONS
MuMag=1.0 ! Permeabilities:
MuAir=1.0
Mu=MuAir ! default to Air
TH = 0.4
MzMag = 10000 ! permanent magnet strength
Mz = 0 ! global magnetization variable
Nx = vector(0,Mz,0)
Ny = vector(-Mz,0,0)
B = curl(Ax,Ay,0) ! magnetic induction vector
Bxx= -dz(Ay)
Byy= dz(Ax) ! unfortunately, "By" is a reserved word.
Bzz= dx(Ay)-dy(Ax)
EQUATIONS
Ax: div(grad(Ax)/mu+Nx) = 0
Ay: div(grad(Ay)/mu+Ny) = 0
EXTRUSION
surface 'lower Iron Bottom' z=-20
layer 'lower iron'
SURFACE "Magnet Bottom" Z=-10
LAYER "Magnet"
SURFACE "Magnet Top" Z=-1.1
Layer 'Air Bottom'
SURFACE 'Iron Bottom' z=-1
LAYER "Iron"
Surface 'Iron Top' z=0
BOUNDARIES
surface "lower Iron Bottom" value(Ax)=0 value(Ay)=0
region 1 {Air bounded by conductive box }
start (70,0)
value(Ax)=0 value(Ay)=0
arc(center=0,0) angle=360 to finish
limited region 2
layer 'lower iron' Mu=1000*MuAir
layer "magnet" Mu=MuAir Mz = 0
start (-40,-10) line to (-2,-10) to (-2,10) to (-40,10) to finish
limited region 3
layer 'lower iron' Mu=1000*MuAir
```

```
layer "magnet" Mu=1000*MuAir Mz = 0
start(-40,-10) line to (-30,-10) to (-30,10) to (-40,10) to finish
limited region 4 { Lower Magnet }
layer "Magnet" Mu=MuMag Mz = MzMag
start (-25,-10) line to (-2,-10) to (-2,10) to (-25,10) to finish
limited region 5 { Iron}
layer "Iron" Mu=1000*MuAir
start(25,0) arc(center=0,0) angle=360
start (-40,-10) line to (-30,-10) to (-30,10) to (-40,10) to finish
feature 'r2p3' start(-25,0) line to (-2,0)
MONITORS
grid(y,z) on x=0
grid(x,z) on y=0
grid(x,y) on z=0
contour(log10(Mu)) on y=0 painted
contour(Ax) zoom(-60,-60,120,120) on x=0
contour(Ay) zoom(-60,-60,120,120) on y=0
PLOTS
grid(y,z) on x=0 contour(log10(Mu)) on x=z zoom(-40,-20,40,40) painted
grid(x,z) on y=0 contour(log10(Mu)) on x=y painted zoom(-40,-20,40,40)
grid(x,y) on z=0 contour(log10(Mu)) on y=0 painted zoom(-40,-20,40,40)
contour(Ax) zoom(-40,-20,40,40) on x=0
contour(Ay) zoom(-40,-20,40,40) on y=0
vector(Byy,Bzz) zoom(-40,-20,40,40)on x=0 norm
vector(Bxx,Bzz) zoom(-26,-17,26,26) on y=0 norm
contour(Bzz) zoom(-40,-40,80,80) on z=0
elevation(Bzz) from(-2,0) to (-25,0) on z=0
END
```

H Script for solving magnetohydrodynamics in an insulating channel

```
{The magnetic induction applied in the y direction is
   0<B0<818.34 gauss. Because of the size of the channel
   there results a very magnetohydrodynamic flow at
   818.34 gauss or 0.081834 Tesla. Here the total flow is
   kept at an average value of 0.1 meter/second in the
   the channel to produce laminar flow. }
   TITLE 'MHD Channel with Insulating Walls'
   COORDINATES
   CARTESIAN( 'z' , 'y' )
   VARIABLES
   u
   H ! u in M/sec, H in Amp/M and is th x component of
```

```
! magnetic field intensity
GLOBAL VARIABLES
Pg ! A global variable must be used to make the total flow constant
! Pg is calculated in the Equations Section
SELECT
errlim = 1e-4
stages = 8
DEFINITIONS
sigma = 2.21E6 ! at 200 degrees C.
mu = 5.92E-4 ! viscosity in Nwt-sec/m^2
rho = 507 ! density in Kg/m^3
Jz = -dy(H) Jy = dz(H)
J = vector(Jz,Jy)
M = staged(0,1,1.291549665,4.641588834,10,12.91549665,46.41588837,100)
B0 = M/(0.02*((sigma/mu)^0.5))
E = (Jz-B0*u)/sigma emf = - integral(xcomp(E))
Fmag = - Jz*B0 Re = rho*0.1*0.02/mu
EQUATIONS
u:del2(u) +(B0/mu)*Dy(H) = -Pg/mu
H:del2(H) + sigma*B0*Dy(u) = 0
Pg: integral(u)=0.1 ! Here the total flow is set equal to 0.1 cubic meters per second
BOUNDARIES
REGION 1
start(0,-0.01) value(u)=0 value(H) = 0 line to (0.01,-0.01) to
(0.01,0.01) to (0,0.01) natural(H)=0 natural(U)=0 line to finish
MONITORS
contour(u) report(M) report(B0) report(Pg)
! surface(u) as 'Velocity Surface' report(M) report(B0)
! surface(H) as 'Magnetic Intensity Surface' report(M) report(B0)
PLOTS
contour(u) as 'Velocity Profile in m/sec' report(M) report(B0) report(Pg) report(Re)
contour(H) as 'Current Flow Paths in amps/m' report(M) report(B0) report(Pg)
surface(u) as 'Velocity Surface in m/sec' report(M) report(B0) report(Pg)
surface(H) as 'Magnetic Intensity Surface in amps/m' report(M) report(B0) report(Pg)
elevation(E) as 'Electric Field on z axis in volt/m' from (0,0) to (0.01,0) report(M) report(B0) report(Pg)
vector(J) as 'Current Density in amps/m^2' report(M) report(B0) report(Pg)
vector(E) as 'Electric Field in volts/m' report(M) report(B0) report(Pg)
vector(J) zoom(0,-0.01,0.01,0.001) as 'Current Density in amps/m^2' report(M) report(B0) report(Pg)
vector(E) zoom(0,-0.01,0.01,0.001) as 'Electric Field in volts/m' report(M) report(B0) report(Pg)
```

elevation(Jz) as 'Current Density in amps/m^2' from (0, 0) to (0,0.01) report(M) report(B0) report(Pg)
contour(Fmag) as 'Induced Magnetic Force in x direction' report(M) report(B0) report(Pg)
history(Pg) as '- Pressure gradient versus Hartmann number' versus M report(M) report(B0) report(Pg)
END

I Script for sea water flowing in an insulating pipe

{The magnetic induction applied in the y direction is
6.28<B0<125,511 gauss. This is accomplished by a strong
superconducting magnet. Here the total flow is kept
constant in channel to produce laminar flow for Re=2500. }
TITLE 'MHD Pipe with Insulating Walls'
COORDINATES
CARTESIAN('z' , 'y')
VARIABLES
u
H ! u in M/sec, H in Amp/M and is th x component of
! magnetic field intensity
GLOBAL VARIABLES
Pg ! A global variable must be used to make the total flow constant
! Pg is calculated in the Equations Section
SELECT
errlim = 1e-4
stages = 16
DEFINITIONS
sigma = 4.8! sea water
mu = 1E-3 ! viscosity in Nwt-sec/m^2
rho = 1.025E-3 ! density in Kg/m^3
nu = 9.8E-7 ! kinematic viscosity
D = 0.23 ! pipe inside diameter in meters
Re = 2500 ! Reynolds number to avoid turbulence
Q = nu*Pi*D*Re/4 ! calculated flow in m^3/sec
Jz = -dy(H) Jy = dz(H)
J = vector(Jz,Jy)
M = staged(0.001,0.5,1,1.5,2,2.5,3,3.5,4,5,6,7,8,10,15,20)
B0 = M/(0.023*((sigma/mu)^0.5))
E = (Jz-B0*u)/sigma emf = - integral(xcomp(E))
Fmag = - Jz*B0
EQUATIONS
u:del2(u) +(B0/mu)*Dy(H) = -Pg/mu

```
H:del2(H) + sigma*B0*Dy(u) = 0
Pg: integral(u) = Q ! Here the total flow is set equal to Q cubic meters per second
BOUNDARIES
REGION 1
start(0,0.115) line to (0,0) to (0.115,0)
value(u) = 0 value(H) = 0 arc(center=0,0) angle = 90
line to finish
MONITORS
contour(u) report(M) report(B0) report(Pg) report(Q)
! surface(u) as 'Velocity Surface' report(M) report(B0)
! surface(H) as 'Magnetic Intensity Surface' report(M) report(B0)
PLOTS
contour(u) as 'Velocity Profile in m/sec' report(M) report(B0) report(Pg) report(Q)
contour(H) as 'Current Flow Paths in amps/m' report(M) report(B0) report(Pg)
surface(u) as 'Velocity Surface in m/sec' report(M) report(B0) report(Pg) report(Q)
surface(H) as 'Magnetic Intensity Surface in amps/m' report(M) report(B0) report(Pg) report(Q)
elevation(E) as 'Electric Field on z axis in volt/m' from (0,0) to (0.115,0) report(M) report(B0) report(Pg)
vector(J) as 'Current Density in amps/m^2' report(M) report(B0) report(Pg)
vector(E) as 'Electric Field in volts/m' report(M) report(B0) report(Pg) Report (Q)
elevation(Jz) as 'Current Density in amps/m^2' from (0, 0) to (0,0.115) report(M) report(B0) report(Pg)
contour(Fmag) as 'Induced Magnetic Force in x direction' report(M) report(B0) report(Pg)
history(Pg) as '- Pressure gradient versus Hartmann number' versus M report(M) report(B0) report(Pg)
END
```

I.1 Discussion of results

Below are three plots of the many that are generated by the above script. The Hartmann number is five requiring 31.4 kilogauss for the magnetic induction applied in the y direction. This pipe is the size of a blue whale's aorta whose blood electrical is of the order of magnitude of sea water depending upon his red blood cell count. Magnetic inductions of this size are only in the vicinity of super conductors. Fig.103 and Fig.104 show the velocity profile and the magnetic intensity. Fig.105 exhibits the electric field for y = 0 as a function of z. It also gives $\int_{z=0}^{z=0.115} \mathbf{E} \cdot d\ell = -3.135 \times 10^{-3}$ volts. Thus the emf is given by minus twice that or 6.27 millivolts.

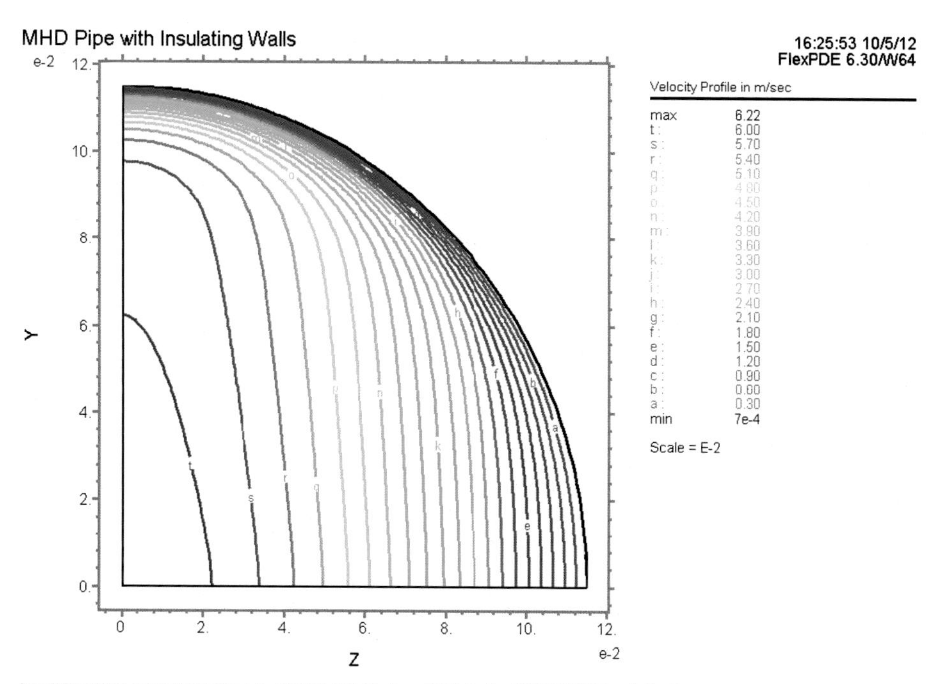

Figure 103: Velocity profile for M = 5

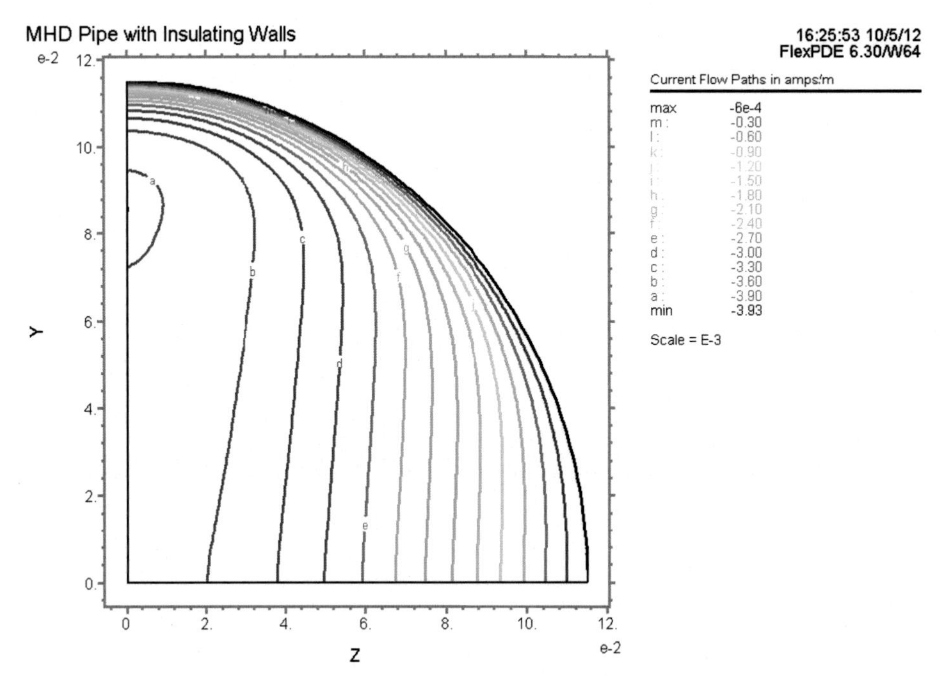

Figure 104: Magnetic intensity for M = 5

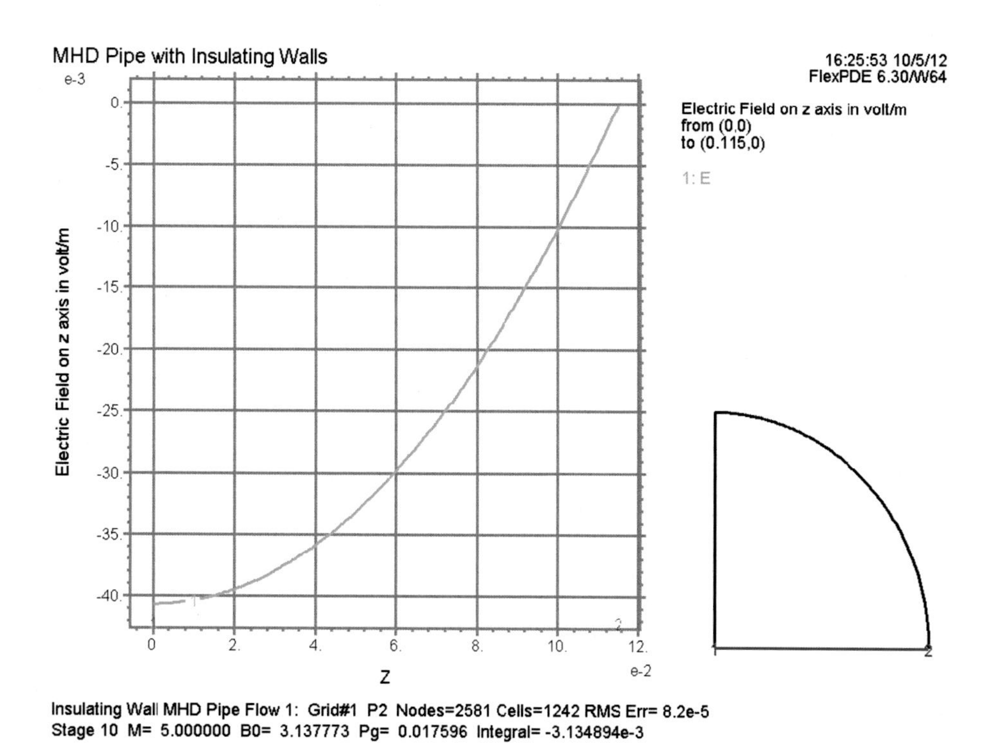

Figure 105: Electric field on z axis

J Script for sea water flowing in an ideal conducting pipe

```
{The magnetic induction applied in the y direction is
  6.28<B0<125,511 gauss. This is accomplished by a strong
  superconducting magnet. Here the total flow is kept
  constant in channel to produce laminar flow for Re=2500. }
  TITLE 'MHD Pipe with Conducting Walls'
  COORDINATES
  CARTESIAN( 'z' , 'y' )
  VARIABLES
  u
  H ! u in M/sec, H in Amp/M and is th x component of
  ! magnetic field intensity
  GLOBAL VARIABLES
  Pg ! A global variable must be used to make the total flow constant
  ! Pg is calculated in the Equations Section
  SELECT
  errlim = 1e-4
  stages = 10
  DEFINITIONS
  sigma = 4.8! sea water
  mu = 1E-3 ! viscosity in Nwt-sec/m^2
  rho = 1.025E-3 ! density in Kg/m^3
  nu = 9.8E-7 ! kinematic viscosity
  D = 0.23 ! pipe inside diameter in meters
  Re = 2500 ! Reynolds number to avoid turbulence
  Q = nu*Pi*D*Re/4 ! calculated flow in m^3/sec
  Jz = -dy(H) Jy = dz(H)
  J = vector(Jz,Jy)
  M = staged(0.001,0.5,1,1.5,2,2.5,3,3.5,4,5)
  B0 = M/(0.023*((sigma/mu)^0.5))
  E = (Jz-B0*u)/sigma emf = - integral(xcomp(E))
  Fmag = - Jz*B0
  EQUATIONS
  u:del2(u) +(B0/mu)*Dy(H) = -Pg/mu
  H:del2(H) + sigma*B0*Dy(u) = 0
  Pg: integral(u) = Q ! Here the total flow is set equal to Q cubic meters per
second
  BOUNDARIES
  REGION 1
  start(0,0.115) value(u) = 0 arc(center=0,0) angle = 360
  line to finish
  MONITORS
  contour(u) report(M) report(B0) report(Pg) report(Q)
```

! surface(u) as 'Velocity Surface' report(M) report(B0)
!surface(H) as 'Magnetic Intensity Surface' report(M) report(B0)
PLOTS
contour(u) as 'Velocity Profile in m/sec' report(M) report(B0) report(Pg) report(Q)
contour(H) as 'Current Flow Paths in amps/m' report(M) report(B0) report(Pg)
surface(u) as 'Velocity Surface in m/sec' report(M) report(B0) report(Pg) report(Q)
surface(H) as 'Magnetic Intensity Surface in amps/m' report(M) report(B0) report(Pg) report(Q)
elevation(E) as 'Electric Field on z axis in volt/m' from (-0.115,0) to (0.115,0) report(M) report(B0) report(Pg)
vector(J) as 'Current Density in amps/m^2' report(M) report(B0) report(Pg)
vector(E) as 'Electric Field in volts/m' report(M) report(B0) report(Pg) Report (Q)
elevation(Jz) as 'Current Density in amps/m^2' from (0,-0.115) to (0,0.115) report(M) report(B0) report(Pg)
contour(Fmag) as 'Induced Magnetic Force in x direction' report(M) report(B0) report(Pg)
history(Pg) as '- Pressure gradient versus Hartmann number' versus M report(M) report(B0) report(Pg)
END

The boundaries section is very simple because the whole pipe cross section is used and the omission of a boundary condition on H defaults to $\partial \mathbf{H}/\partial \mathbf{n} = 0$,the desired condition. Although the solution yields many plots not exhibited here, a few are given. In Fig.106 and Fig.107 are shown the velocity profile and induced current vectors for a small Hartmann number. The velocity profile is not influenced much and the induced current is very small. However, its path to the boundaries verifies the boundary condition used for H. When the Hartmann number is five Fig.108 and Fig.109 show the typical flattening of the velocity profile and the increased value of the current vectors. In Fig.110 the electric field along the z axis is shown.

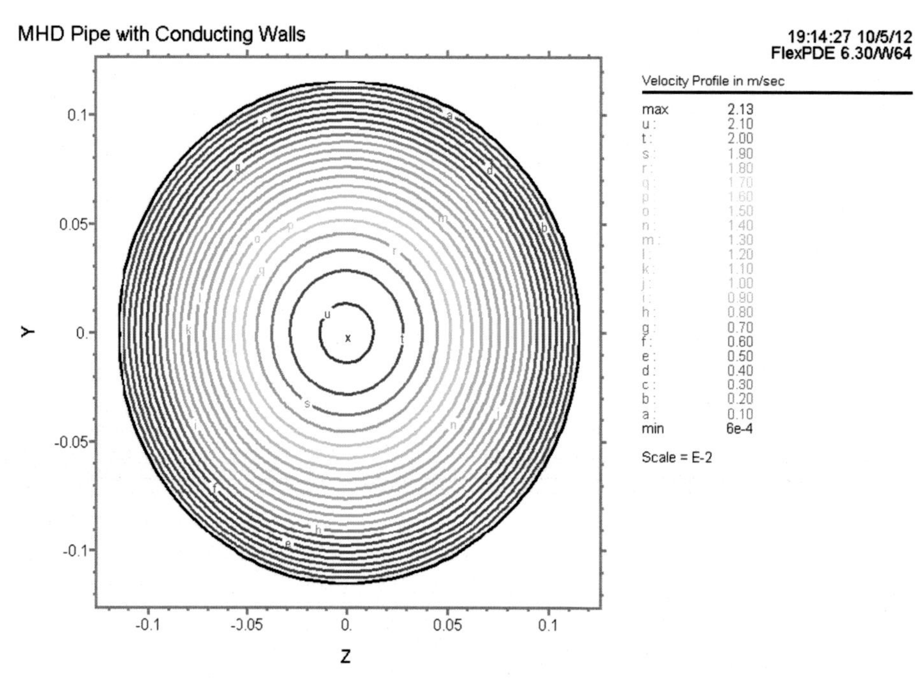

Figure 106: Velocity profile at $M \approx 0$

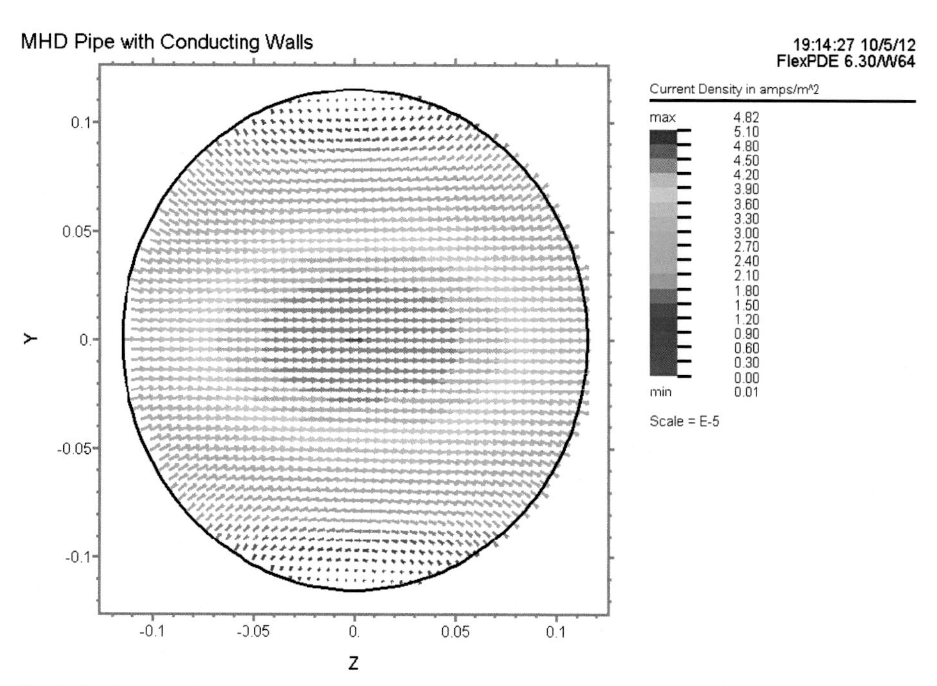

Figure 107: Current vectors at $M \approx 0$

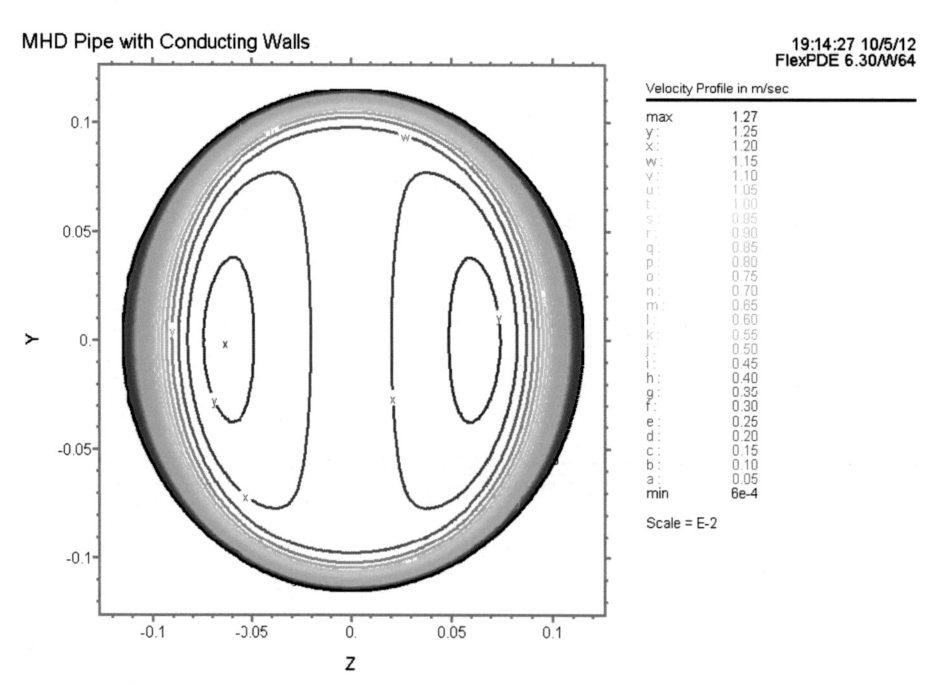

Conducting Wall MHD Pipe Flow 2: Grid#2 P2 Nodes=21645 Cells=10618 RMS Err= 5.9e-5
Stage 10 M= 5.000000 B0= 3.137773 Pg= 0.532735 Q= 4.425719e-4 Integral= 4.424348e-4

Figure 108: Velocity profile for $M = 5$

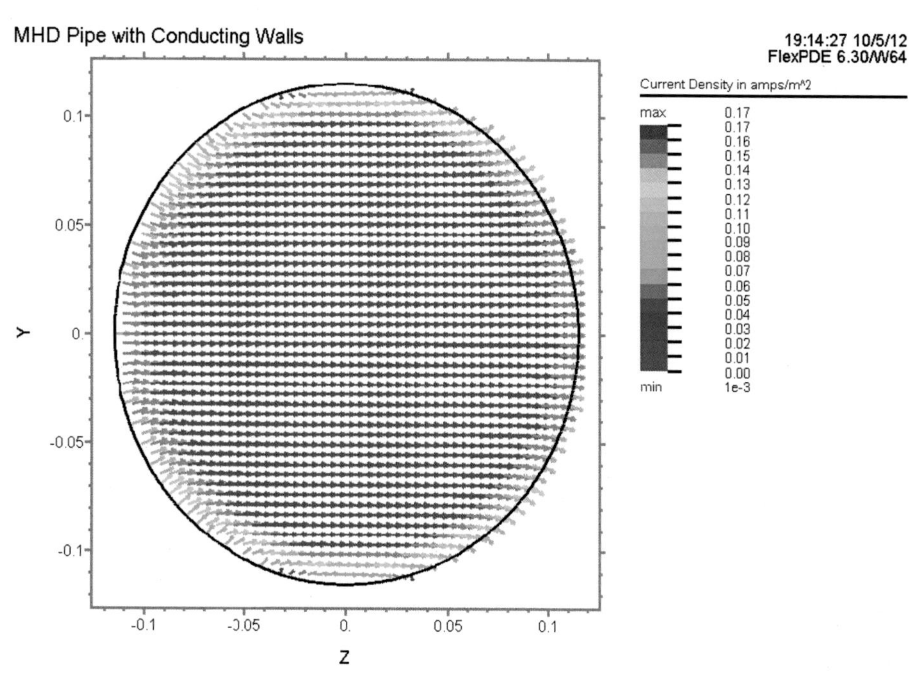

Figure 109: Current vectors for $M = 5$

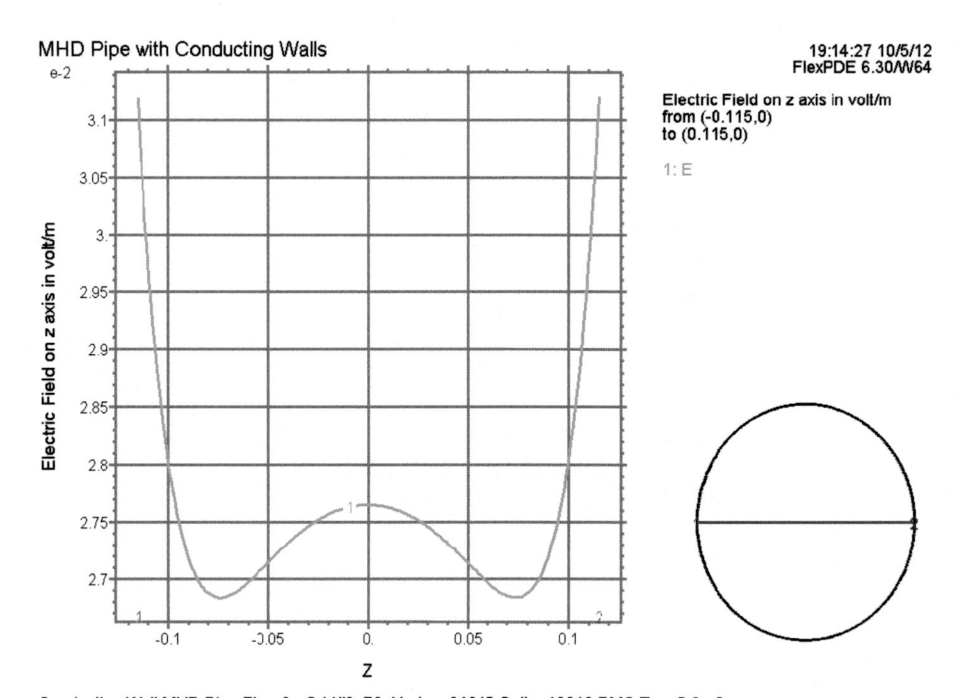

Figure 110: Electric field on the z axis for $M = 5$

K Script for problem 9.5.3 with ideal conducting walls

```
{The magnetic induction applied in the y direction is
  0<B0<818.34 gauss. Because of the size of the channel
  there results a very magnetohydrodynamic flow at
  818.34 gauss or 0.081834 Tesla. The total flow is adjusted
  so that Re = 2500 in the channel to produce laminar flow. }
  TITLE 'MHD Channel with Ideal Conducting Walls'
  COORDINATES
  CARTESIAN( 'z' , 'y' )
  VARIABLES
  u
  H ! u in M/sec, H in Amp/M and is th x component of
  ! magnetic field intensity
  GLOBAL VARIABLES
  Pg ! A global variable must be used to make the total flow constant
  ! Pg is calculated in the Equations Section
  Select
  stages = 11
  errlim = 1E-4
  DEFINITIONS
  mu = 5.92E-4 ! viscosity in Nwt-sec/m^2
  rho = 507 ! density in Kg/m^3
  nu = 9.8E-7 ! kinematic viscosity
  D = 0.23 ! pipe inside diameter in meters
  Re = 2500 ! Reynolds number to avoid turbulence
  Q = 0.02*nu*Re ! calculated flow in m^3/sec
  sigma = 2.21E6 ! liquid lithium at 200 degrees C.
  Jz = -dy(H) Jy = dz(H)
  J = vector(Jz,Jy) E = curl(H)
  M = staged(0,1,1.3,5,10,13,30,50,63,75,100)
  B0 = M/(0.02*((sigma/mu)^0.5))
  Ez = (Jz-B0*u)/sigma
  Fmag = - Jz*B0 ! Magnetic retarding force
  EQUATIONS
  u:del2(u) +(B0/mu)*Dy(H) = -Pg/mu
  H:del2(H) + sigma*B0*Dy(u) = 0
  Pg: integral(u) = Q ! Here the total flow adjusted so that Re = 2500
  BOUNDARIES
  REGION 1
  start(0,-0.01) value(u)=0 line to (0.01,-0.01) to
  (0.01,0.01) to (0,0.01) natural(U)=0 line to finish
  MONITORS
  contour(u) report(M) report(B0) report(Pg) Report (Q)
```

! surface(u) as 'Velocity Surface' report(M) report(B0)
! surface(H) as 'Magnetic Intensity Surface' report(M) report(B0)
PLOTS
contour(u) as 'Velocity Profile in m/sec' report(M) report(B0) report(Pg) report(Q)
contour(H) as 'Current Flow Paths in amps/m' report(M) report(B0) report(Pg)
surface(u) as 'Velocity Surface in m/sec' report(M) report(B0) report(Pg)
surface(H) as 'Magnetic Intensity Surface in amps/m' report(M) report(B0) report(Pg)
elevation(E) as 'Electric Field on z axis in volt/m' from (0,0) to (0.01,0) report(M) report(B0) report(Pg)
vector(J) as 'Current Density in amps/m^2' report(M) report(B0) report(Pg)
vector(E) as 'Electric Field in vlots/m' report(M) report(B0) report(Pg)
vector(J) zoom(0,-0.01,0.01,0.001) as 'Current Density in amps/m^2' report(M) report(B0) report(Pg)
vector(E) zoom(0,-0.01,0.01,0.001) as 'Electric Field in volts/m' report(M) report(B0) report(Pg)
elevation(Jz) as 'Current Density in amps/m^2' from (0, 0) to (0,0.01) report(M) report(B0) report(Pg)
contour(Fmag) as 'Induced Magnetic Force in x direction' report(M) report(B0) report(Pg)
history(Pg) as '- Pressure gradient versus Hartmann number' versus M report(M) report(B0) report(Pg) Report (Q)
END

In the above script the right half of the geometry is used by making symmetry about the y axis. Fig.111shows the velocity profile present when there is no magnetic induction applied in the y direction. Fig.112 shows the velocity profile for $M = 100$, a rather large and highly magnetohydrodynamic value. The core of the lithium travels almost like a solid conductor. In Fig.113 the current vectors are exhibited and are pointing in the z direction except near the corners. The electric field in the z direction along $y = 0$ is given in Fig.114. The integral given there must be doubled to get the induced emf. Fig.115 shows the z component of current density along the y axis. It is almost constant except at the right edge. Fig.116 exhibits the history of the calculation of pressure gradient necessary to keep the Reynolds number at 2500 to avoid turbulence in the flow.

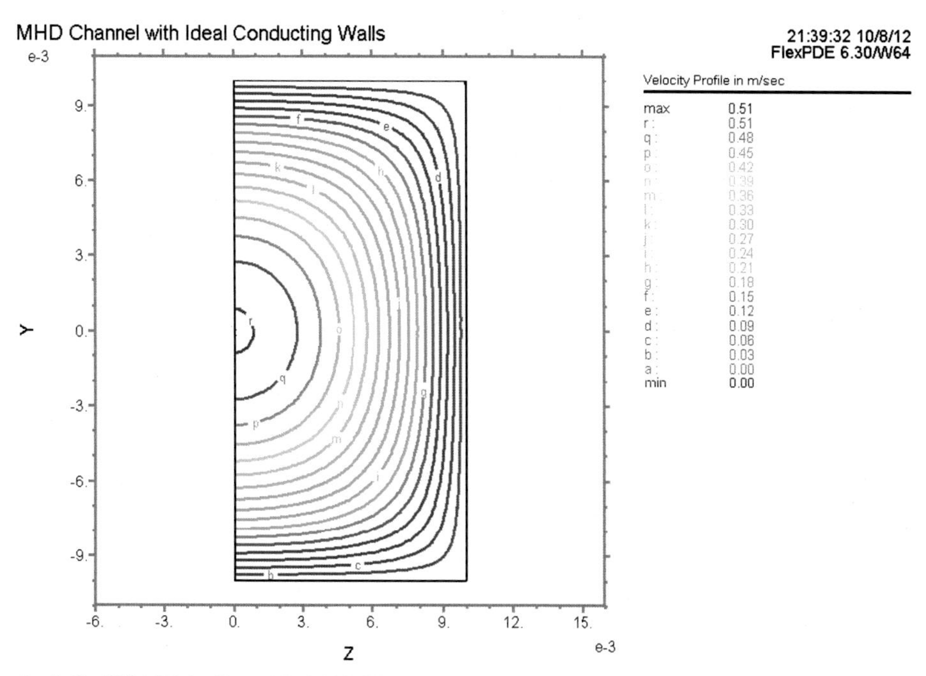

Figure 111: Velocity profile for $M = 0$

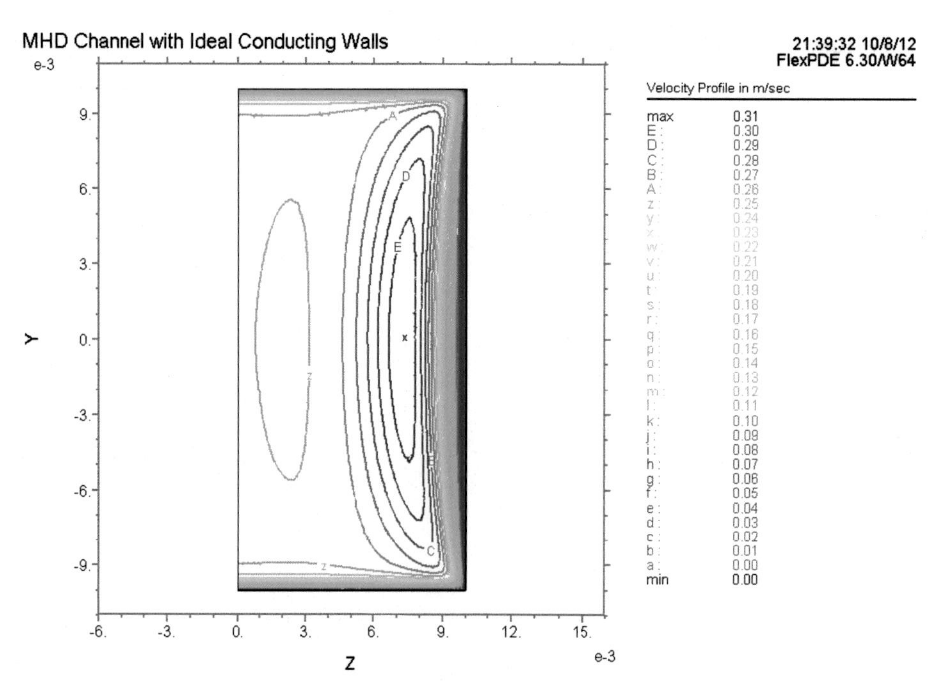

Figure 112: Velocity profile for $M = 100$

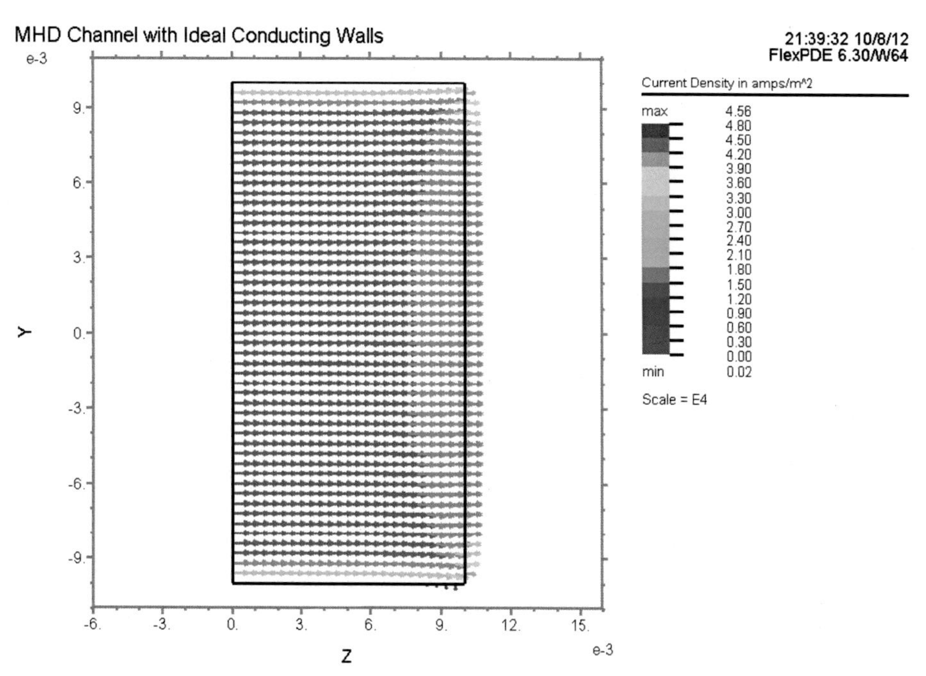

Figure 113: Current vectors for $M = 100$

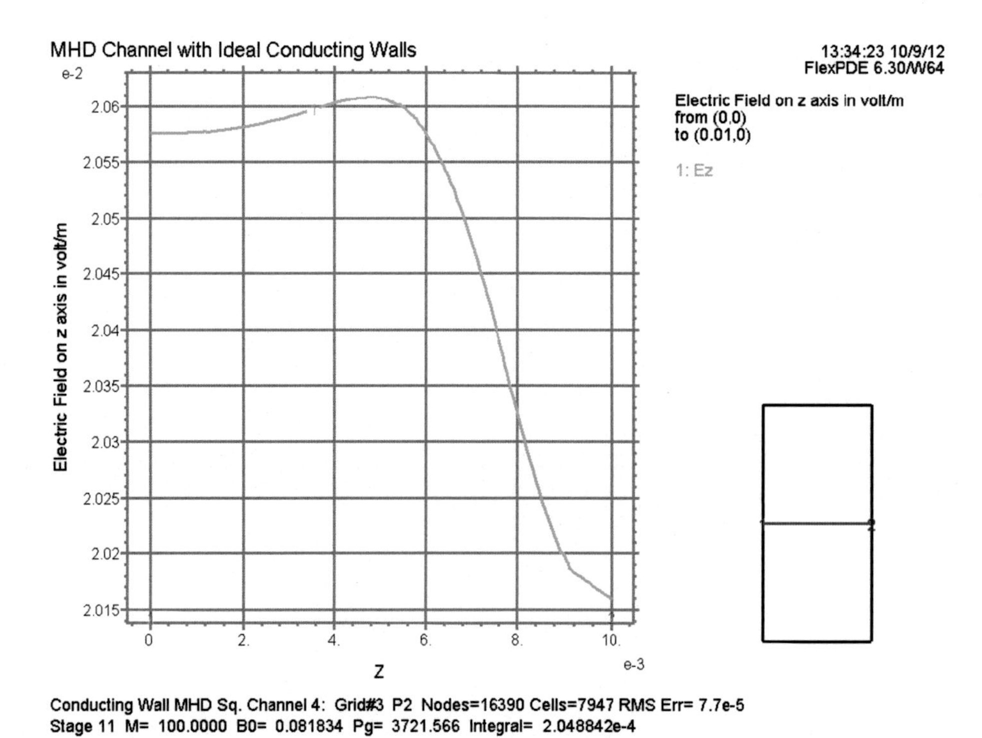

Figure 114: Electric field, E_z on the z axis for $M = 100$

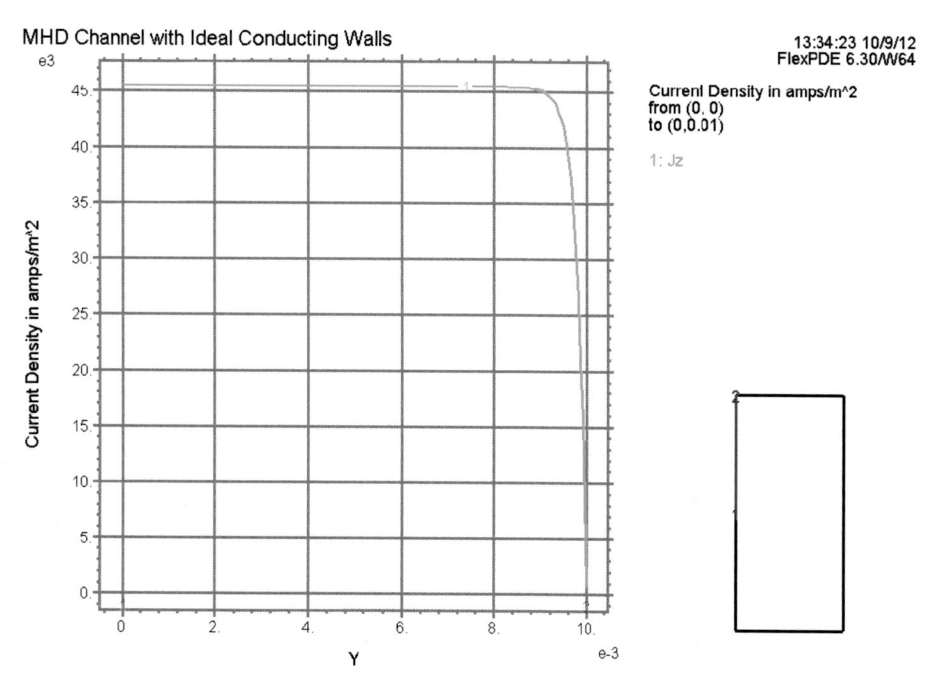

Figure 115: Current density, J_z on the z axis

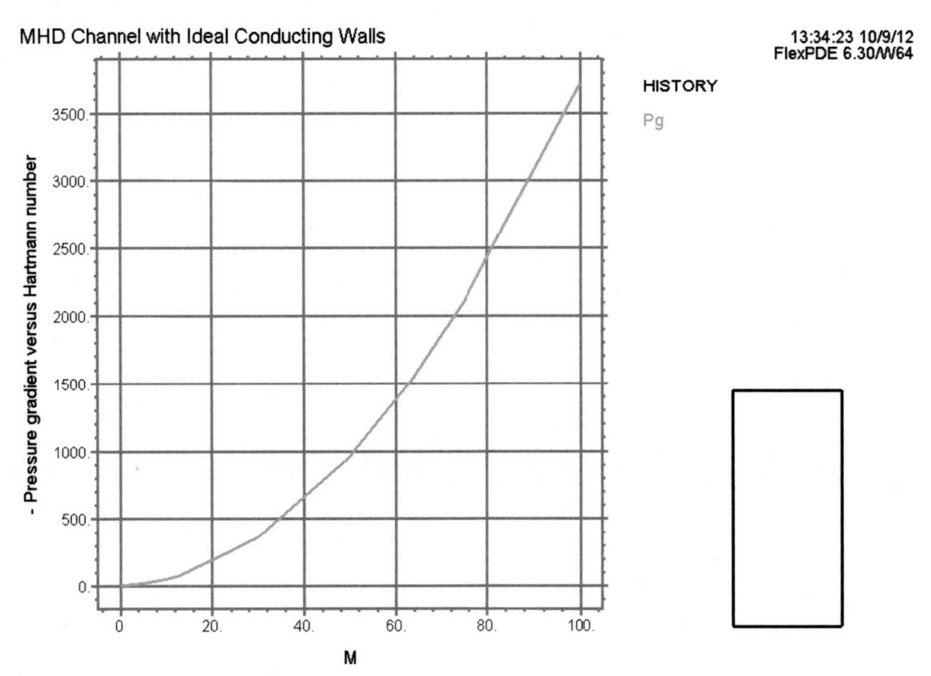

Figure 116: Negative pressure gradient for M = 100

L Script for sea water flowing in a half-full pipe

L.1 Ideally insulating wall

```
{The magnetic induction applied in the y direction is
   6.28<B0<125,511 gauss. This is accomplished by a strong
   superconducting magnet. Here the total flow is kept
   constant in channel to produce laminar flow for Re=2500.
   Only the right half is needed due to symmetry about the xy plane }
   TITLE 'Half-full MHD Pipe with Insulating Wall'
   COORDINATES
   CARTESIAN( 'z' , 'y' )
   VARIABLES
   u
   H ! u in M/sec, H in Amp/M and is th x component of
   ! magnetic field intensity
   GLOBAL VARIABLES
   Pg ! A global variable must be used to make the total flow constant
   ! Pg is calculated in the Equations Section
   SELECT
   errlim = 1e-4
   stages = 16
   DEFINITIONS
   sigma = 4.8! sea water
   mu = 1E-3 ! viscosity in Nwt-sec/m^2
   rho = 1.025E-3 ! density in Kg/m^3
   nu = 9.8E-7 ! kinematic viscosity
   D = 0.23 ! pipe inside diameter in meters
   Re = 2500 ! Reynolds number to avoid turbulence
   Q = nu*Pi*D*Re/4 ! calculated flow in m^3/sec
   Jz = -dy(H) Jy = dz(H)
   J = vector(Jz,Jy)
   M = staged(0.001,0.5,1,1.5,2,2.5,3,3.5,4,5,6,7,8,10,15,20)
   B0 = M/(0.023*((sigma/mu)^0.5))
   Ez = u*B0+Jz/sigma
   Fmag = - Jz*B0
   EQUATIONS
   u:del2(u) +(B0/mu)*Dy(H) = -Pg/mu
   H:del2(H) + sigma*B0*Dy(u) = 0
   Pg: integral(u) = Q ! Here the total flow is set equal to Q cubic meters per
second
   BOUNDARIES
   REGION 1
   start(0.115,0.115) value(H)=0 line to (0,0.115) natural(H)=0 line to (0,0)
   value(u) = 0 value(H) = 0 arc(center=0,0.115) angle = 90
   line to finish
```

MONITORS
contour(u) report(M) report(B0) report(Pg) report(Q)
! surface(u) as 'Velocity Surface' report(M) report(B0)
! surface(H) as 'Magnetic Intensity Surface' report(M) report(B0)
PLOTS
contour(u) as 'Velocity Profile in m/sec' report(M) report(B0) report(Pg) report(Q)
elevation(u) as 'Free Surface Velocity in m/sec' from (0,0.115) to (0.115,0.115) report(M) report(B0) report(Pg)
contour(H) as 'Current Flow Paths in amps/m' report(M) report(B0) report(Pg)
!surface(u) as 'Velocity Surface in m/sec' report(M) report(B0) report(Pg) report(Q)
!surface(H) as 'Magnetic Intensity Surface in amps/m' report(M) report(B0) report(Pg) report(Q)
elevation(Ez) as 'Electric Field on z axis in volt/m' from (0,0.115) to (0.115,0.115) report(M) report(B0) report(Pg)
vector(J) as 'Current Density in amps/m^2' report(M) report(B0) report(Pg)
elevation(Jz) as 'Current Density in amps/m^2' from (0, 0) to (0,0.115) report(M) report(B0) report(Pg)
contour(Fmag) as 'Induced Magnetic Force in x direction' report(M) report(B0) report(Pg)
history(Pg) as '- Pressure gradient versus Hartmann number' versus M report(M) report(B0) report(Pg)
END

Discussion From the comparison of Fig.117 and Fig.118 it is clear that raising the Hartmann number changes the velocity profile greatly. In Fig.119 and Fig.120 the velocity on the free surface is exhibited. The increase in Hartmann number has caused a bit of flatening and reduced the $\int_{z=0}^{z=0.115} u dz$ slightly even though the total flow had been kept constant by raising the pressure gradient. Camparison of Fig.121 and Fig.122 shows that the current density near the bottom of the insulating pipe has increased greatly with the Hartmann number. Fig.123 shows the increase in pressure gradient needed to keep the total flow constant as M increases. Fig 124 portrays the almost linear rise of emf with the increasing Hartmann number. In Fig.125 the error that would happen if the BLV formula were used in calculating the emf is displayed. To even use this erroneous formula the complete field solution given here would have to be known so that $B_0 \int_{z=0}^{z=0.115} u dz$ could be calculated. This was done in order to find the results given in Fig.125. The error is large for low Hartmann numbers and decreases with increasing Hartmann number. However, a flow meter that would use the BLV formula would be very inaccurate.

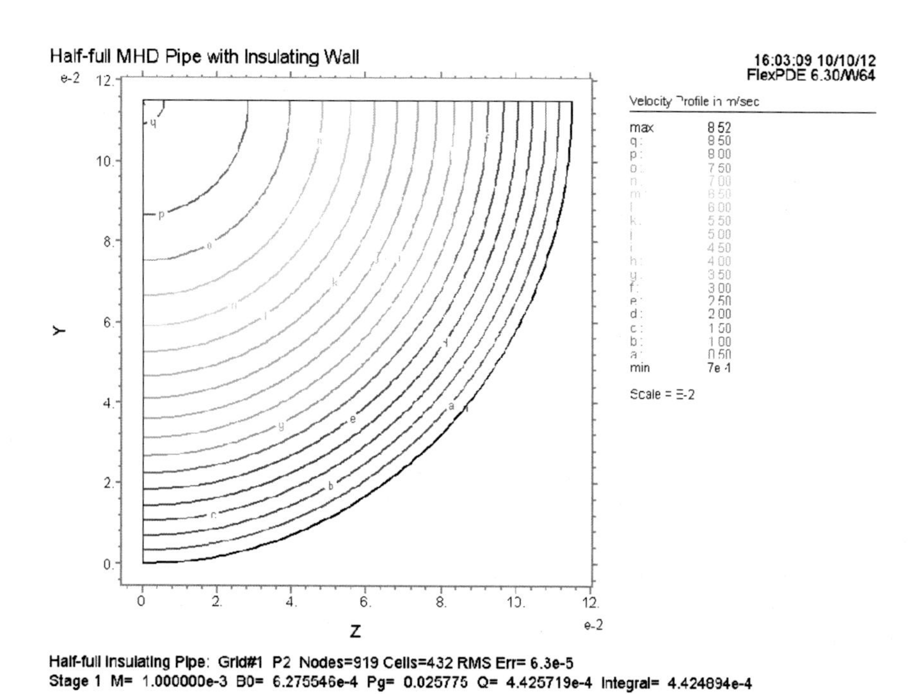

Figure 117: Velocity profile for $M = 0.001$

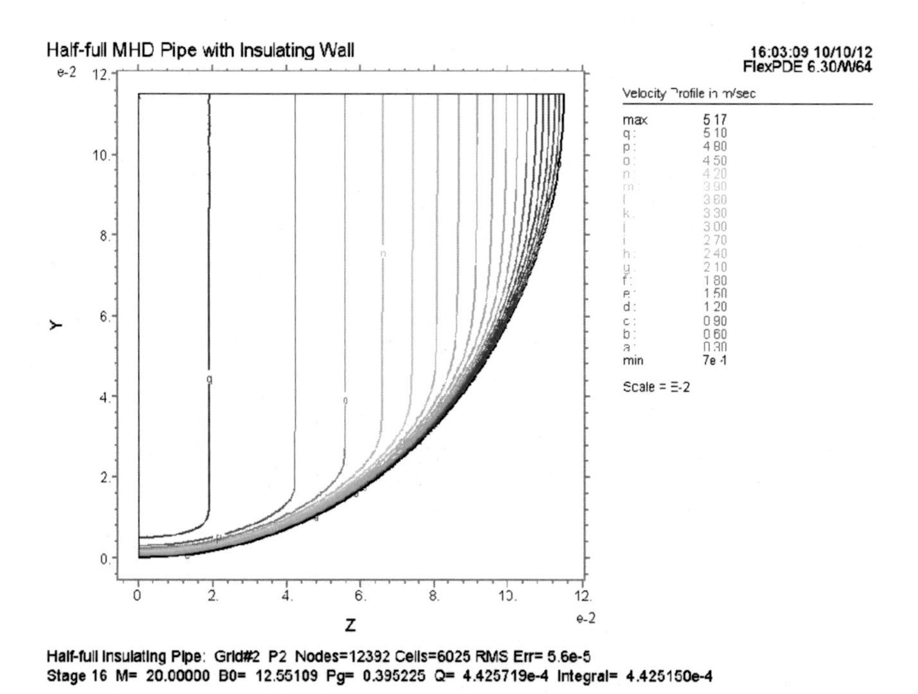

Figure 118: Velocity profile for $M = 20$

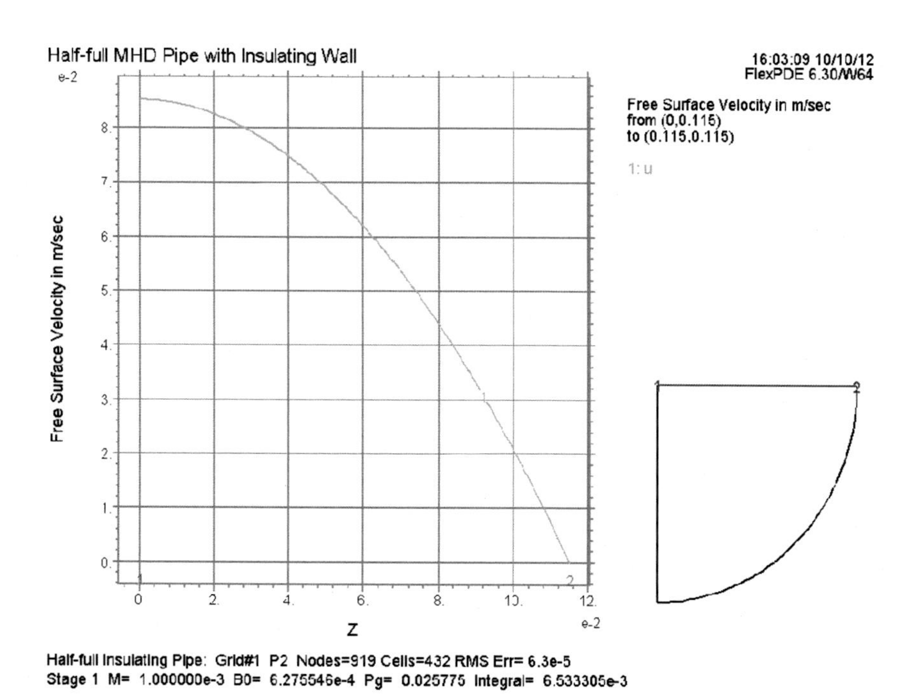

Figure 119: Free surface velocity for $M = 0.001$

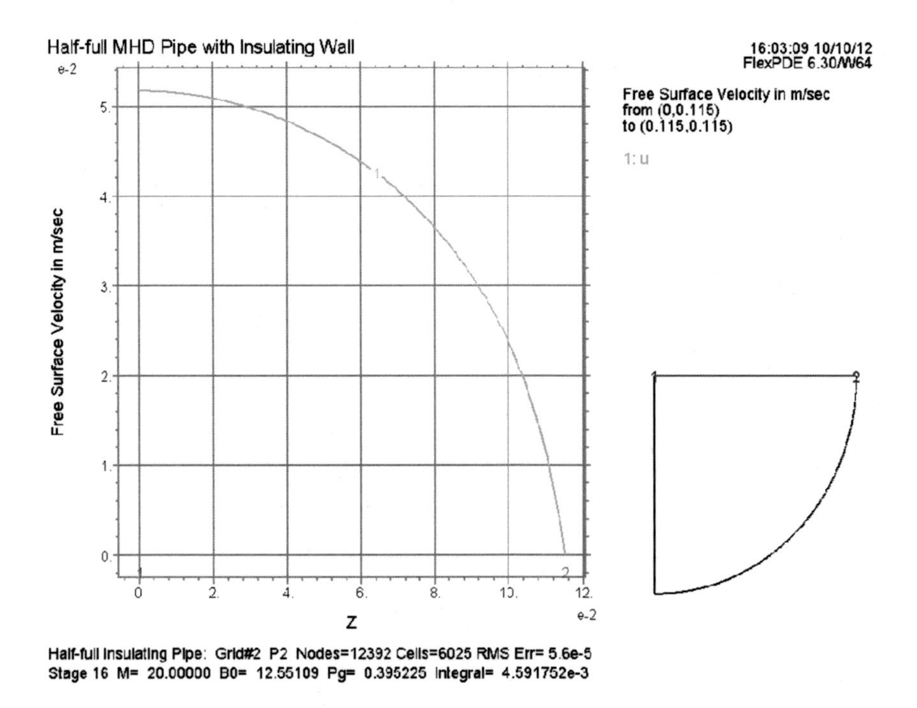

Figure 120: Free surface velocity for $M = 20$

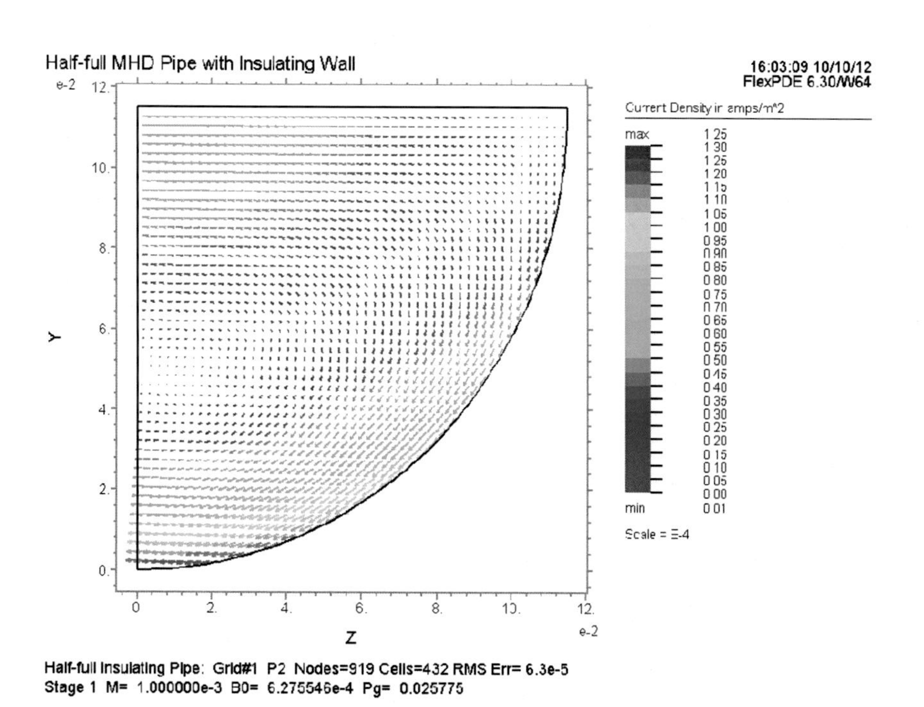

Figure 121: Current density vectors for $M = 0.001$

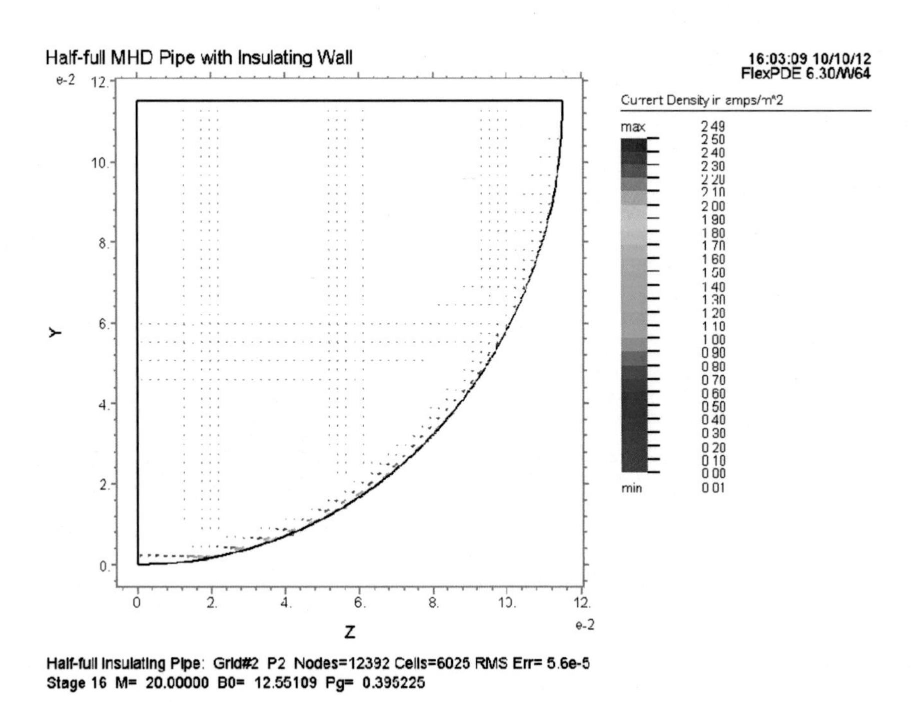

Figure 122: Current density vectors for $M = 20$

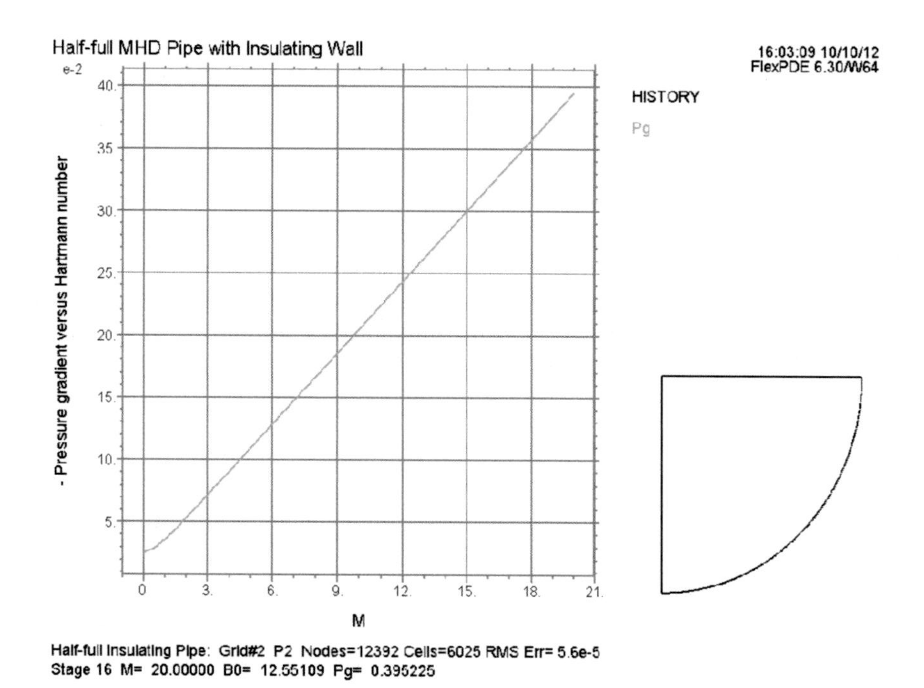

Figure 123: Negative pressure gradient versus M

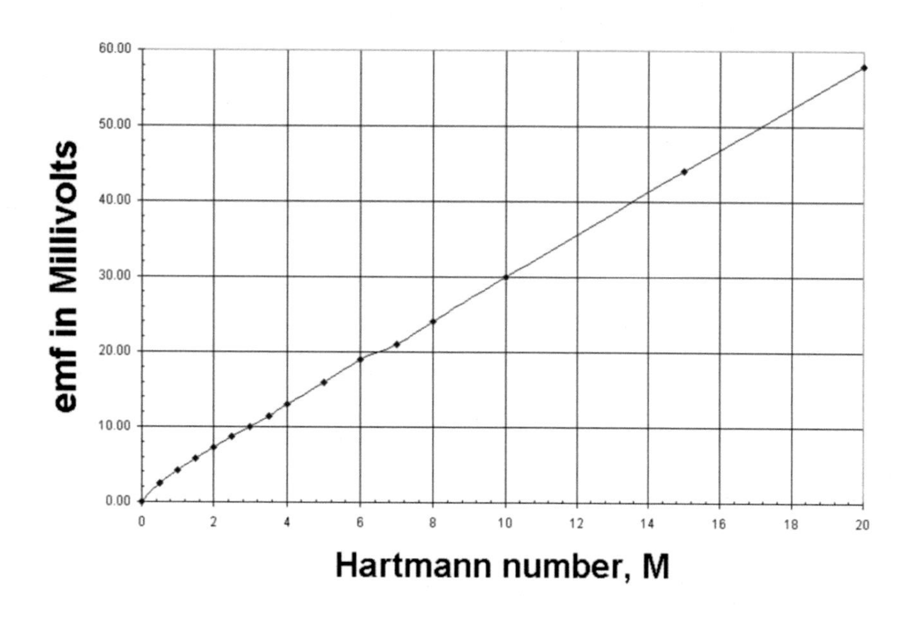

Figure 124: EMF versus Hartmann number

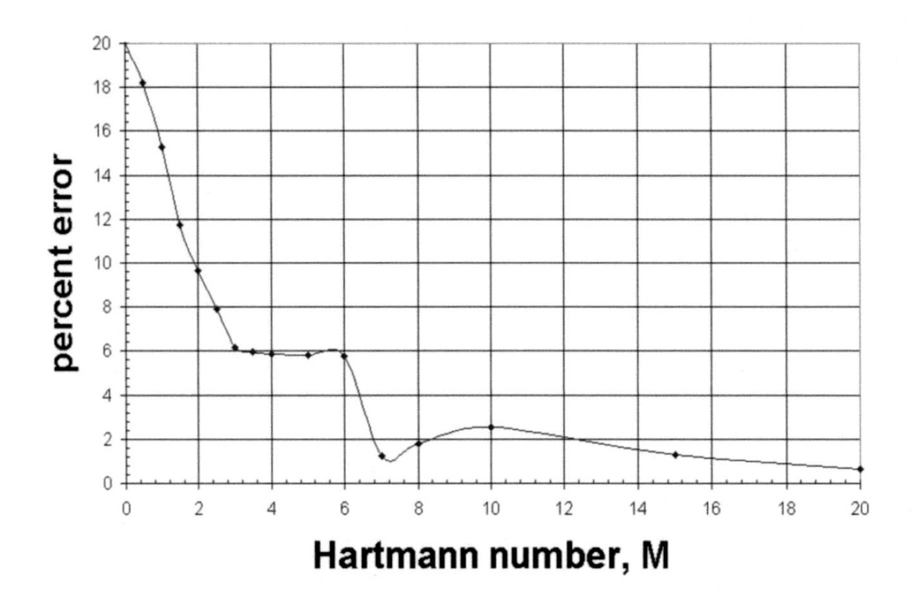

Figure 125: % error in BLV formula versus M

L.2 Ideally conducting wall

L.2.1 Hint

At the ideally conducting wall $\partial H/\partial n = 0$ and to accomplish that the third line of the script given in Appendix L should be changed. In that line natural(H) = 0 should be used instead of value(H) = 0.

L.2.2 Results

Some of the results are presented below. For $M = 0.001$ the velocity profile is the same as given in Fig.117. It follows that the sea water velocities on the free surfaces would also be identical. For $M = 20$ it is interesting to compare Fig.126 to Fig.118. It is clear that the change of pipe material from insulating to conducting drastically changes the velocity profile. Fig.127 shows the free surface velocity for $M = 20$ which is rather different than Fig.120 because of the difference in the velocity profiles of the insulating and conducting pipes. In Fig.128 and 129 clearly the current density vectors and the required pressure gradients are different for the conducting wall pipe than for the insulating wall pipe. In Fig.130 the emf produced with a conducting wall pipe is compared to that obtained with an insulating wall pipe. The error incurred by using the BLV formula to predict the emf is constant, the BLV formula given one-half of the correct answer! If an emf measurement were to be made for these pipes the values given here would have to be doubled because they are for the right half of the geometry. The voltmeter leads in that case would be connected between

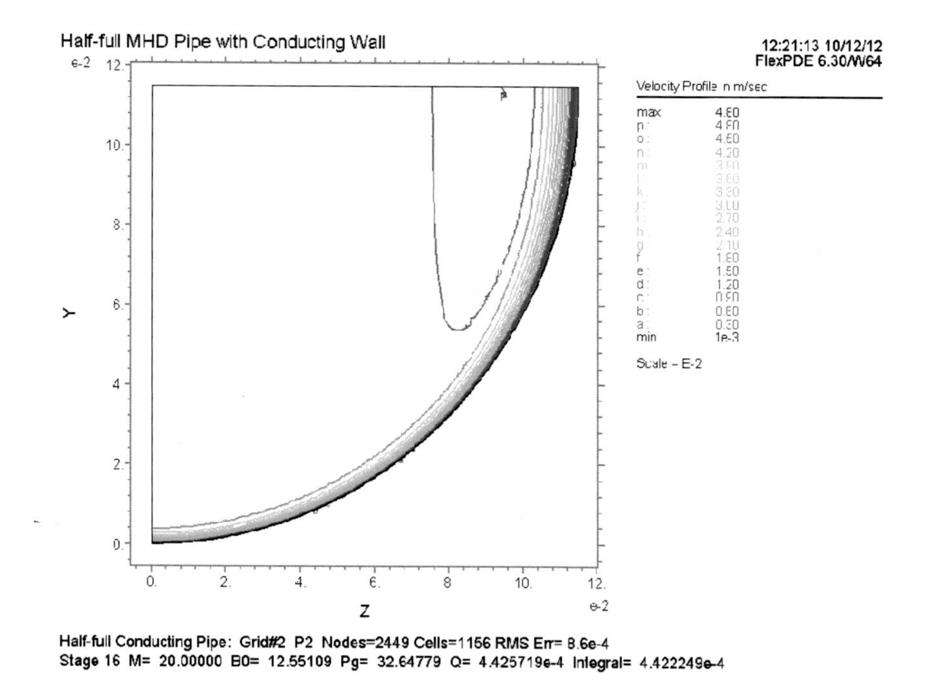

Figure 126: Velocity profile for $M = 20$

points (-0.115,0.115) and (0.115,0.115).

M Script for half-full rectangular channel

```
{The magnetic induction applied in the y direction is
  0<B0<16.37 kilogauss. Because of the size of the channel
  there results a very magnetohydrodynamic flow. It is adjusted
  to have a Reynolds number of 2500 producing laminar flow. }
  TITLE 'MHD Channel with insulating walls & free surface'
  COORDINATES
  CARTESIAN( 'z' , 'y' )
  VARIABLES
  u
  H ! u in M/sec, H in Amp/M and is th x component of
  ! magnetic field intensity
  GLOBAL VARIABLES
  Pg ! A global variable must be used to make the total flow constant
  ! Pg is calculated in the Equations Section
  SELECT
  stages = 11
```

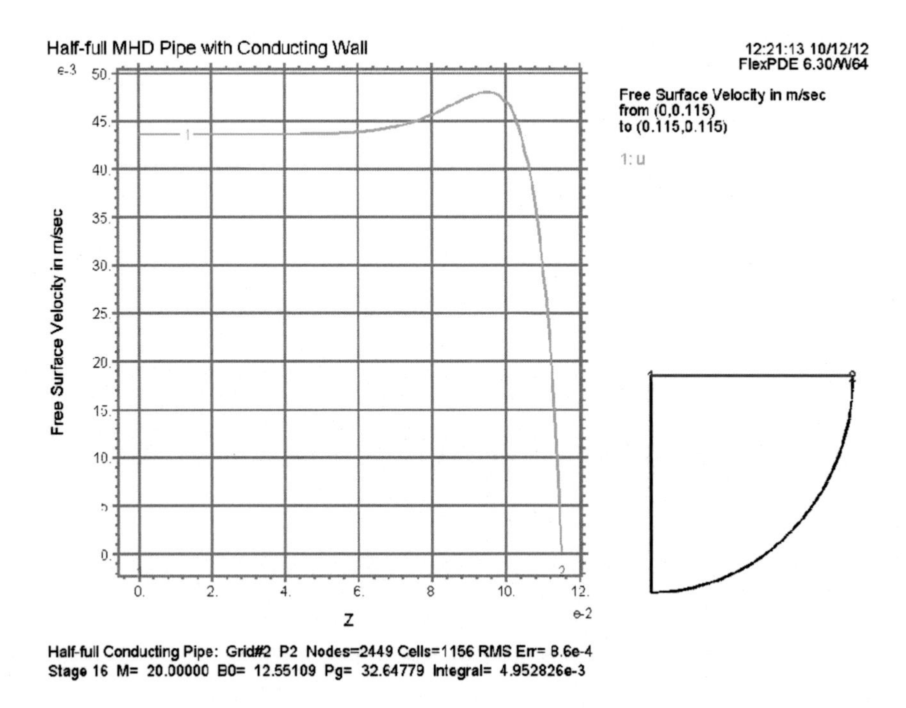

Figure 127: Free surface velocity for $M = 20$

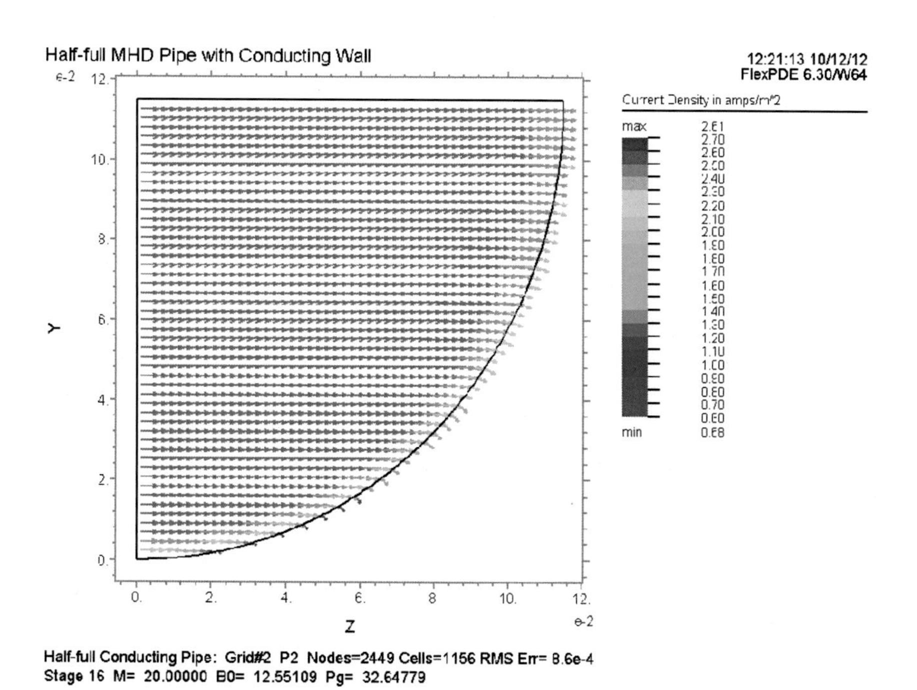

Figure 128: Currebt density vectors $M = 20$

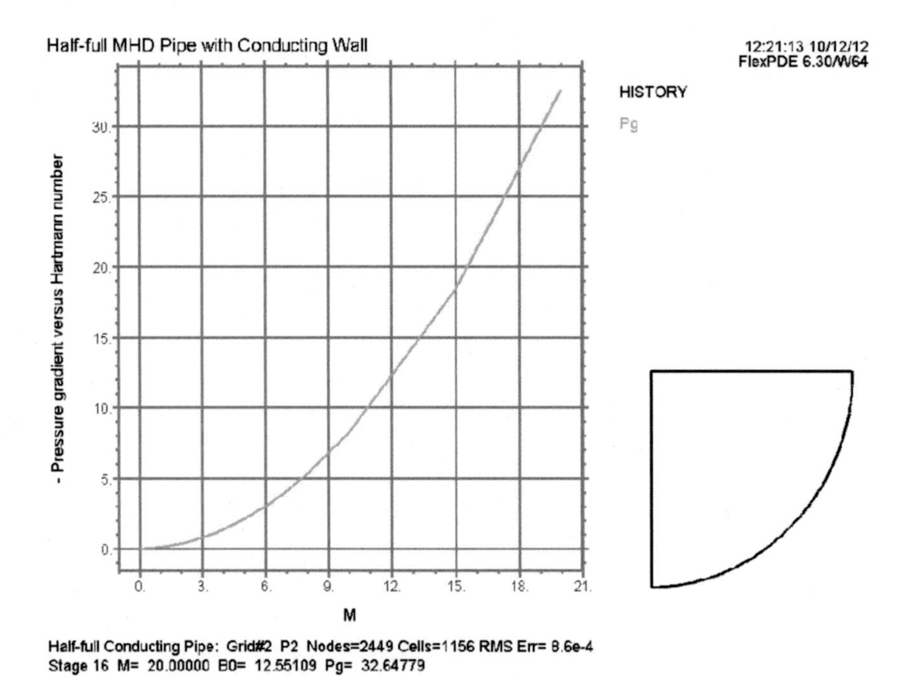

Figure 129: Negative pressure gradient versus M

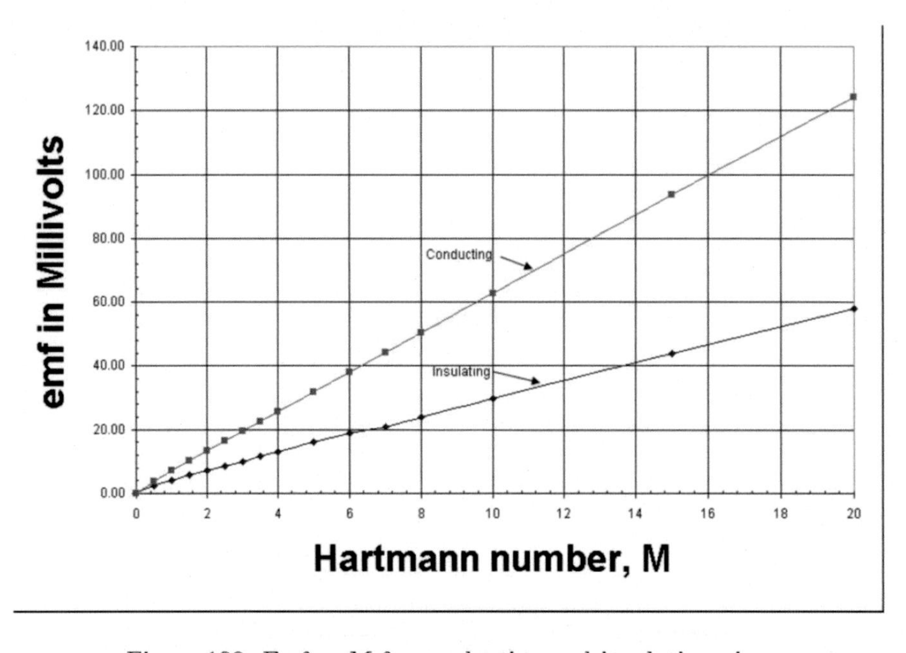

Figure 130: Emf vs M for conducting and insulating pipe

```
errlim = 1E-4
DEFINITIONS
mu = 5.92E-4 ! viscosity in Nwt-sec/m^2
rho = 507 ! density in Kg/m^3
nu = 9.8E-7 ! kinematic viscosity
D = 0.23 ! pipe inside diameter in meters
Re = 2500 ! Reynolds number to avoid turbulence
Q = 0.01*nu*Re ! calculated flow in m^3/sec
sigma = 2.21E6 ! liquid lithium at 200 degrees C.
Jz = -dy(H) Jy = dz(H)
J = vector(Jz,Jy)
M = staged(0,1,1.3,5,10,13,30,50,63,75,100)
B0 = M/(0.01*((sigma/mu)^0.5))
Ez = (Jz-B0*u)/sigma
Fmag = - Jz*B0 ! Magnetic retarding force
EQUATIONS
u:del2(u) +(B0/mu)*Dy(H) = -Pg/mu
H:del2(H) + sigma*B0*Dy(u) = 0
Pg: integral(u) = Q ! Here the total flow adjusted so that Re = 2500
BOUNDARIES
REGION 1
start(0,0) value(u)=0 value(H)=0 line to (0.01,0) to (0.01,0.01)
natural(U)=0 line to (0,0.01) natural(H)=0 line to finish
feature 'Top'
start(0,0.01) line to (0.01,0.01)
MONITORS
contour(u) report(M) report(B0) report(Pg) Report (Q)
PLOTS
contour(u) as 'Velocity Profile in m/sec' report(M) report(B0) report(Pg) report(Q)
contour(H) as 'Current Flow Paths in amps/m' report(M) report(B0) report(Pg)
elevation(u) as 'Velocity on free surface in m/sec' from (0,0.01) to (0.01,0.01) report(M) report(B0) report(Pg)
vector(Ez) as 'Electric Field in volts/m' report(M) report(B0) report(Pg)
elevation(Ez) as 'Electric Field on free surface in volt/m' from (0,0.01) to (0.01,0.01) report(M) report(B0) report(Pg)
vector(Ez) zoom(0,0,0.002,0.002) as 'Electric Field in volts/m' report(M) report(B0) report(Pg)
vector(J) as 'Current Density in amps/m^2' report(M) report(B0) report(Pg)
vector(J) zoom(0,0,0.002,0.002) as 'Current Density in amps/m^2' report(M) report(B0) report(Pg)
contour(Fmag) as 'Induced Magnetic Force in x direction' report(M) report(B0) report(Pg)
history(Pg) as '- Pressure gradient versus Hartmann number' versus M report(M) report(B0) report(Pg) Report (Q)
```

END

M.1 Insulating wall

M.1.1 Results

By symmetry only the right half of the problem need be solved. A Neuman condition on both velocity and magnetic intensity is use along the y axis in the region $0 \leq y \leq 0.01$. Everywhere else $u = H = 0$. The velocity profile for $M = 0$ is given in Fig.131 and for $M = 100$ in Fig.132. As usual the velocity profile is flattened as the Hartmann number increases. The current flow paths which follow the contours of the induced magnetic intensity field are shown in Fig.133 rather than the current density vectors which are hard so see at large Hartmann numbers. In Fig.134 is a plot of free surface velocity versus z at a Hartmann number of one hundred. The surface velocity is almost constant over about 80 percent of the channel width. Fig.135 is a plot of the electric field intensity versus z for $M = 100$. Fig. 136 exhibits a plot of the emf induced along the free surface. The BLV formula differs from the true emf calculated here by orders of magnitude! For that reason no plot is presented.

M.2 Conducting wall

M.2.1 Results

To obtain the correct boundary conditions the two lines following region 1 under Boundaries must be replaced with 'start(0,0) value(u)=0 line to (0.01,0) to (0.01,0.01) natural(U)=0 value(H)=0 line to (0,0.01) natural(H)=0 line to finish'. This keeps the no slip or zero velocity condition on the channel bottom and right side. These conditions make the fluid have a free surface and insures zero magnetic intensity along the free surface allowing no escape of current. Symmetry as mentioned in the insulating case is also preserved. Fig.137 shows the velocity profile for $M = 100$. Notice the difference between it and the corresponding figure for the insulating wall channel. Fig.138 shows current going out the right wall and bottom of the channel. The maximum magnetic intensity is located along the bottom of the channel. With the choice of boundary conditions used all of the current returns under the channel to the left side of the channel on the $y = -0.01$ plane. This could have been altered by interchanging the magnetic boundary conditions on the bottom with those on the top. It is also possible, but more difficult to split the return current between the top and bottom in any desired ratio. The velocity on the free surface shown in Fig.139 shows the same tendencies as did Fig.127 in the case of the half-empty conducting pipe. Fig.140 show the electric field intensity along the free surface. It shows some fluctuations resulting from not making the calculation more accurate by making $Err \approxeq 1 \times 10^{-06}$. However, instead of the calculation running in a few minutes, it might have run for hours.

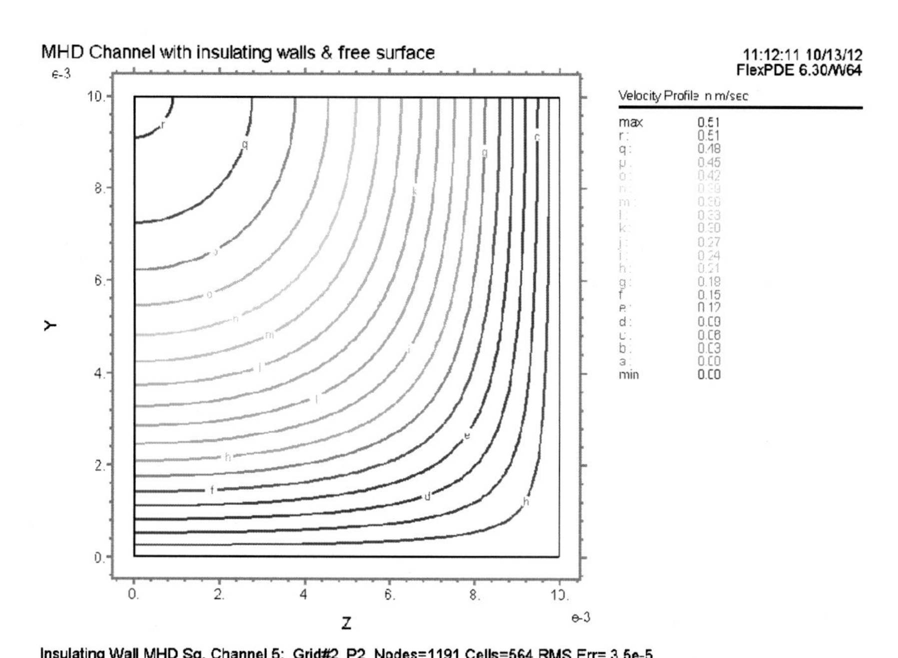

Figure 131: Velocity profile for $M = 0$

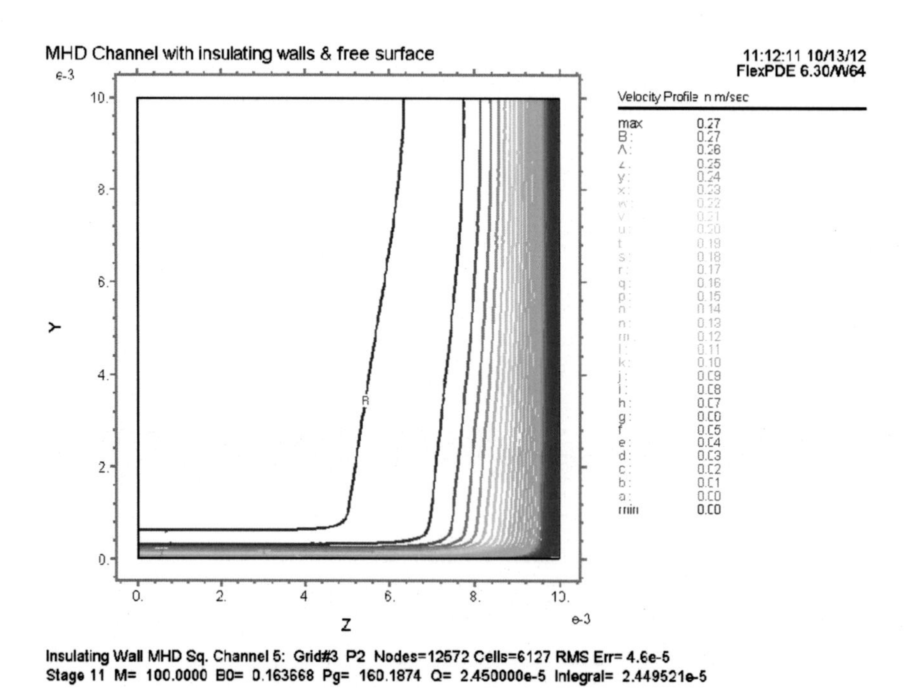

Figure 132: Velocity profile for $M = 100$

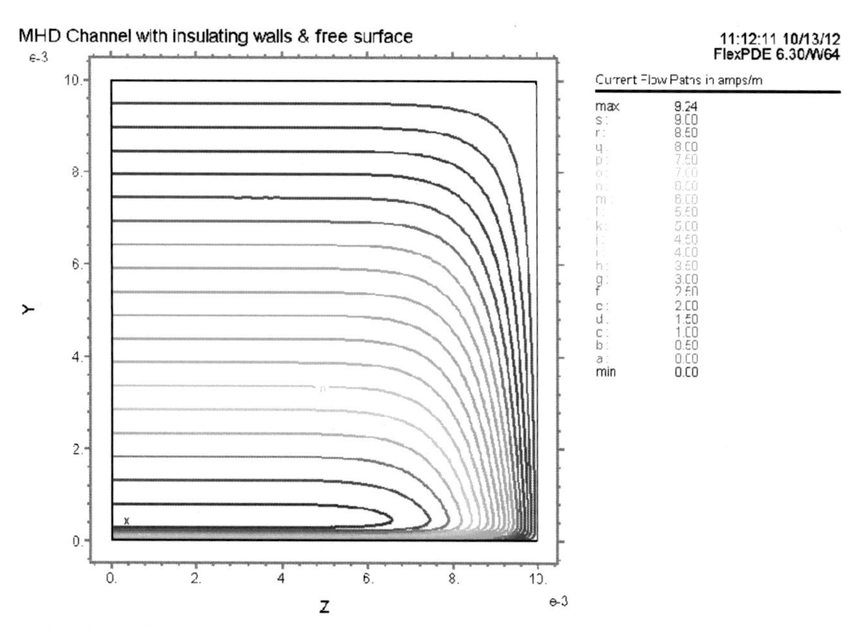

Figure 133: Current flow paths for $M = 100$

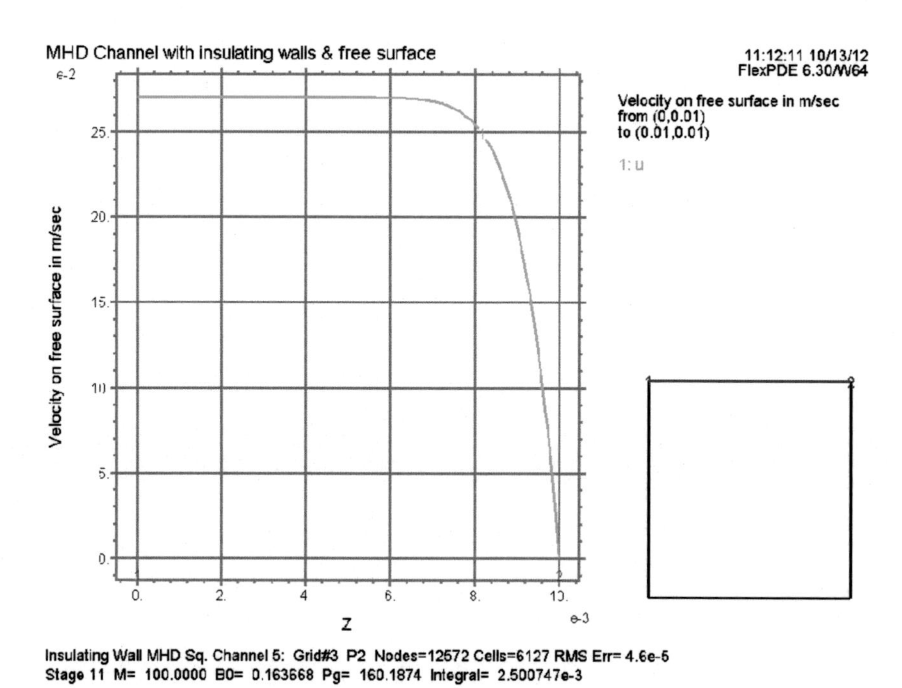

Figure 134: Free surface velocity for $M = 100$

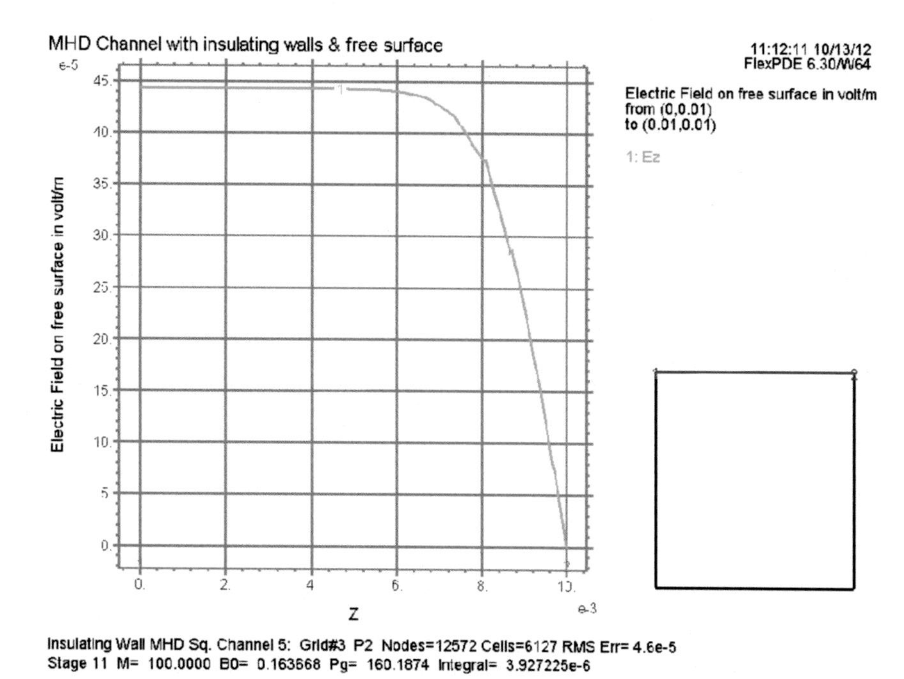

Figure 135: Electric field on free surface for $M = 0$

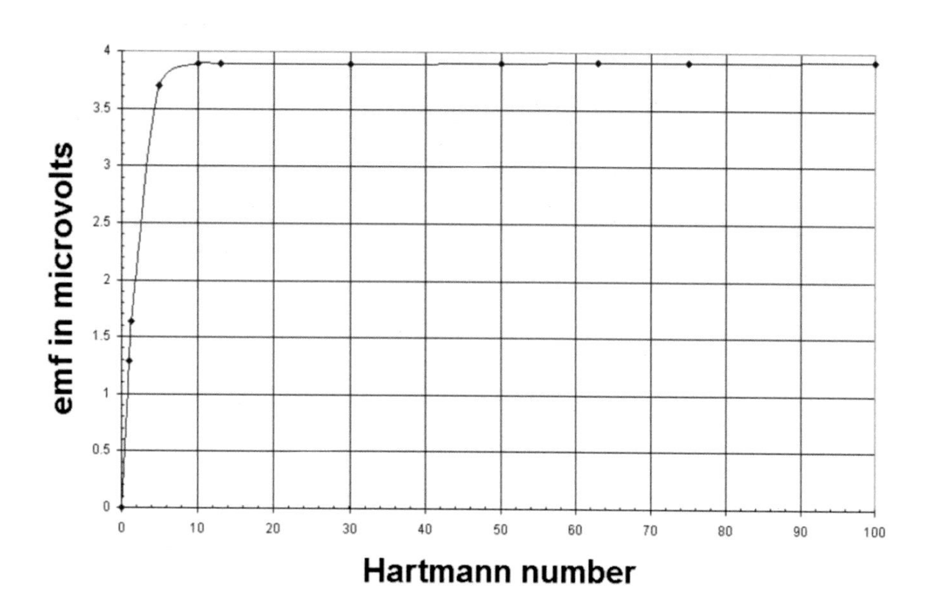

Figure 136: Emf in microvolts

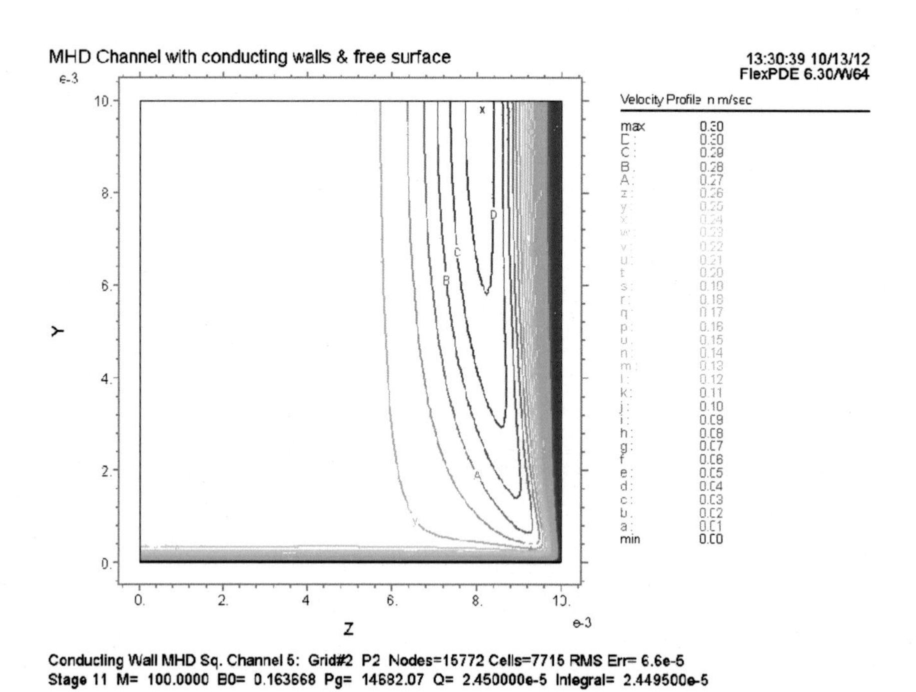

Figure 137: Velocity profile for $M = 100$

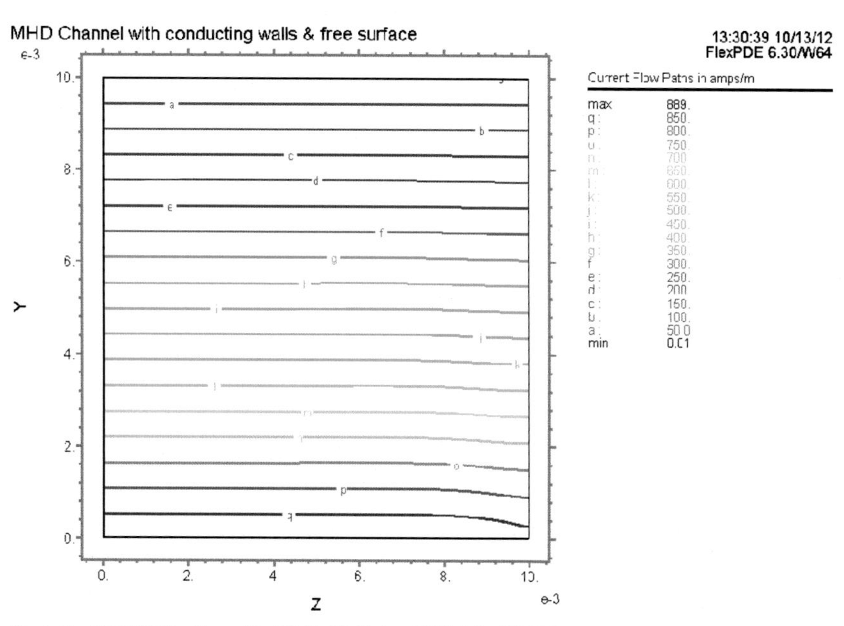

Figure 138: Current flow paths for $M = 100$

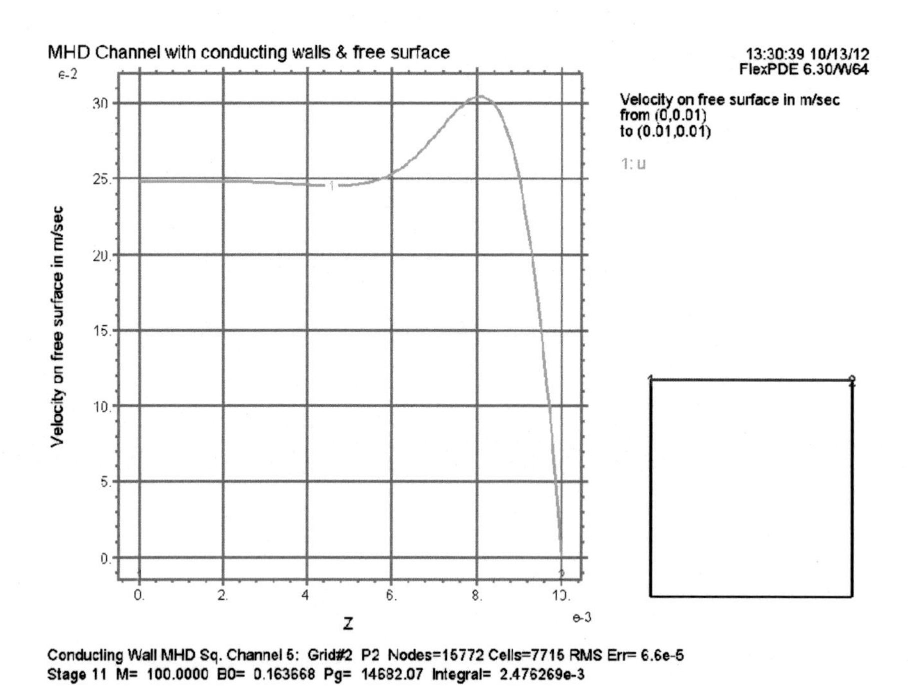

Figure 139: Velocity on free surface for $M = 100$

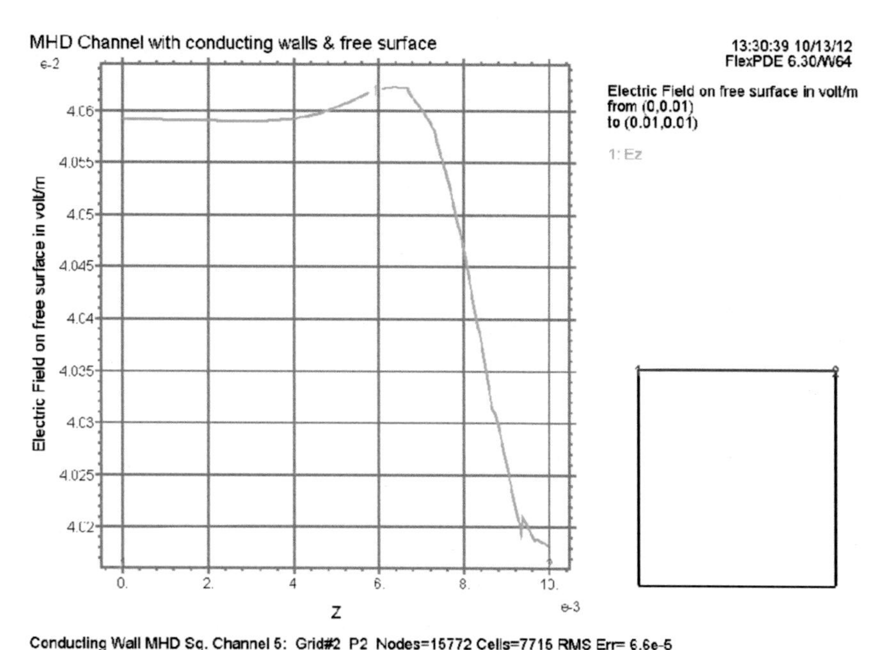

Figure 140: Electric intensity on free surface for $M = 100$

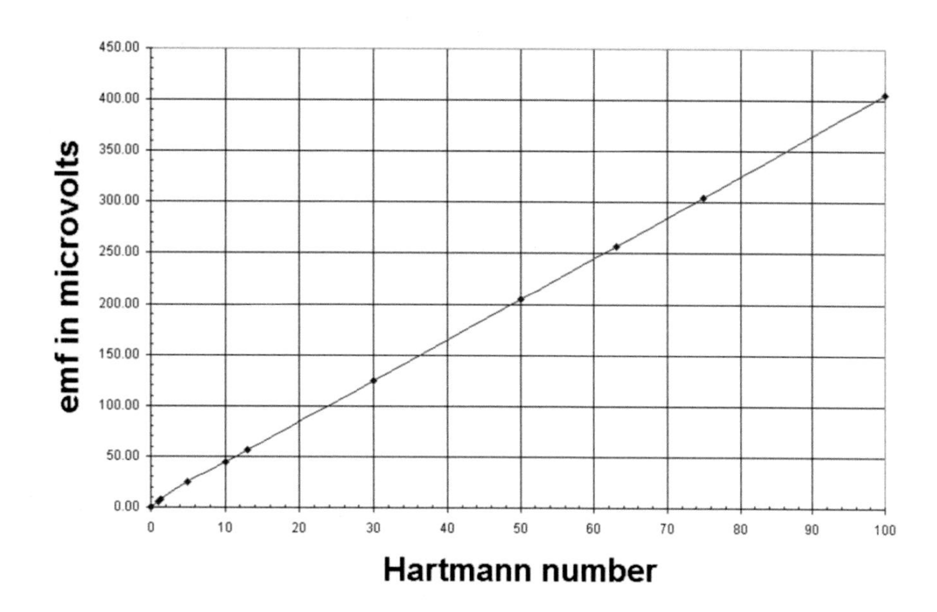

Figure 141: Emf versus Hartmann number

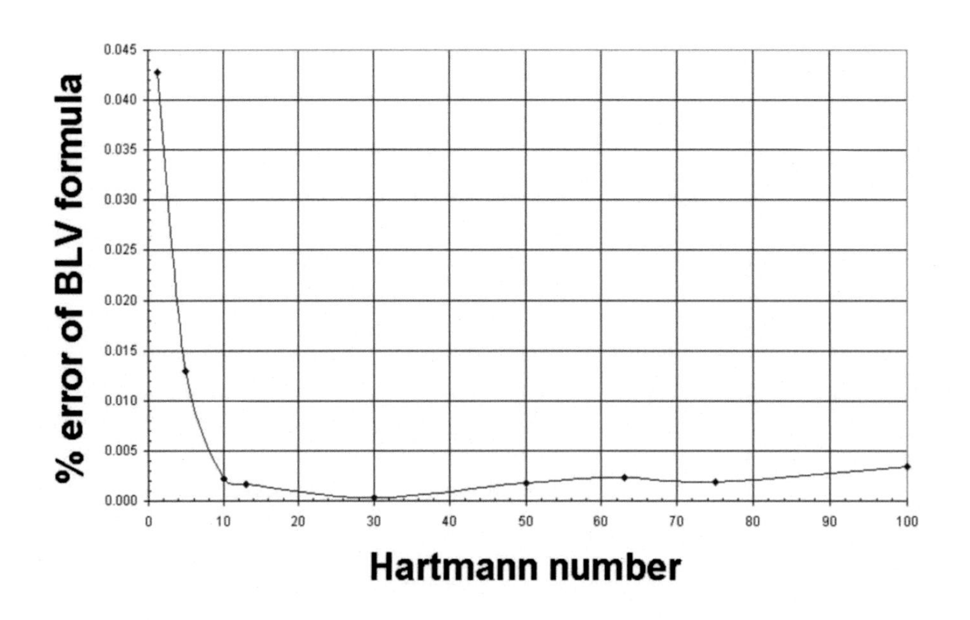

Figure 142: % error in BLV formula

Part VIII

Some of Dr. Joseph Slepian's AIEE "Electrical Essays"

The Transactions of the American Institute of Electrical Engineers were issued monthly beginning in about 1880 and combined with the Institute of Radio Engineers in the 1960s. In the period between 1948 and 1952 the Transactions contained a feature called 'Electrical Essays' written by distinguished electrical engineers. They were designed to pose interesting and challenging problems for readers. The answers were published in later issues. Here we include copies of some of these problems and their answers with the permission of Nancy A. Blair-DeLeon, IEEE Senior Program Manager, Author Support & Content Discoverability, 1-732-562-3965 office telephone, 1-732-439-0813 office mobile, n.blair-deleon@ieee.org (email address). Permission was sought because this author had a rather difficult time finding a library that could get these journals out of storage and it was felt that the readers would encounter difficulty also. The essay given below is related to section 4 of this text entitled 'MAGNETICALLY INDUCED ELECTROMOTIVE FORCE IN RIGID CIRCUITS.'

Interested readers are encouraged to dip two turns of heavy wire into a soap bubble blowing fluid to produce the single sided surface and the two sided surface shown below.

The reader who is used to RMKS units must take care when reading the Electrical Essays because they use the Gaussian system of units that is basically a cgs system in which the permeability and permittivity of vaccuo are unity. In many practical problems this system of units is desirable because centimeters, grams and gauss are very reasonable to describe various engineering devices. However, the units of current and voltage, abamps and abvolts are less desirable.

M.3 Author Comments on the Essays

Study of the essays from the Transactions of American Institute of Electrical Engineers written by Dr. Joseph Slepian illustrates the troubles that arise from not applying the principles of special relativity to the calculation of electromotive force. It is very important to define conductor motion with respect to a fixed but arbitrary observer before attempting to calculate EMF. Although a careful choice of observer can make problem solving easier, all observers see the same EMF in nonrotational magnetic induction fields. The reader shall find the study of the essays especially instructive if he or she works through them in the light of the main body of this book.

M.4 Faraday's Law of Induction II (, July 1949)

Electrical Essays

Faraday's Law of Induction II

A 2-Sided Surface Bounded by 2-Turn Coil

"If S is any smooth, continuous surface bounded by a single, smooth, continuous closed curve, then always the integral of the electric intensity, E, around this closed bounding curve, $\int E_s ds$, is equal to the negative rate of change of the integral over the surface, of the normal component of the magnetic flux density, B.

$$\int E_s ds = -\frac{\partial}{\partial t} \int \int B_n dS$$

"True or false?

"Author's answer: False."

The preceding is an electrical essay which appeared in *Electrical Engineering*, June 1948 (*p 530*).

In his answer the author gave as an example, the Moebius strip illustrated in Figure 1, and observed that such a strip is one-sided and that therefore Faraday's law cannot apply to it, since Faraday's law implies some criterion for determining a positive direction to the normal at any point of the surface, and this cannot be given for a one-sided surface.

The "single, smooth, continuous closed curve" which bounds the Moebius strip is shown in Figure 2 and is nothing more than the curve followed by the conductor of a plain ordinary 2-turn coil! Surely, Faraday's law applies to such a coil.

It then must be possible to construct a 2-sided "smooth, continuous surface," which is bounded by the "smooth, continuous closed curve" followed by the conductor of the 2-turn coil. Given such a surface, we may take, over it, the integral of the normal component of B and thus determine the flux linkages Φ of the coil, to which the equation $E=-\frac{\partial\Phi}{\partial t}$ may be applied. Without such a surface, no meaning can be attached to the idea of flux linkages of the coil.

Question! What does such a 2-sided smooth, continuous surface look like?

J. SLEPIAN (F '27)
(Associate director, Westinghouse Research Laboratories, East Pittsburgh, Pa.)

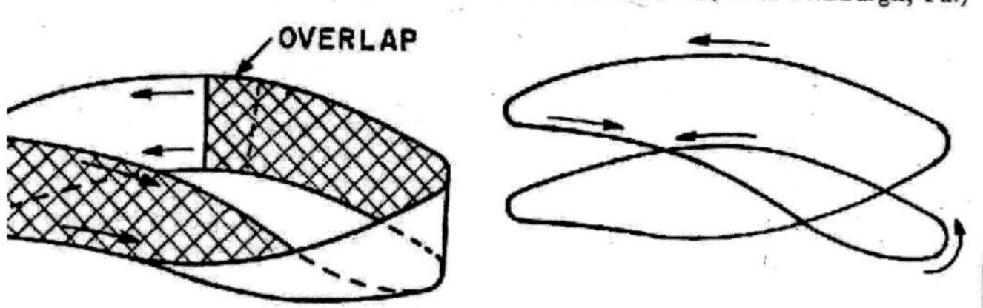

Figure 1 (left). Moebius strip bounded by 2-turn coil
Figure 2 (right). Two-turn coil

Answers to Previous Essays

Faraday's Law of Induction II. The following is the author's answer to a previously published essay of the foregoing title (*EE, Jul '49, p 613*).

The surface in question is shown in Figure 1 of this answer. It is 2-sheeted. Students of the theory of functions of a complex variable will recognize it as portion of a 2-sheeted Riemann surface containing a single branch point. The surface apparently crosses itself at a line of intersection of the two sheets. However, this line of intersection is not to be regarded as common to the two sheets, but the apparent points of this line are to be regarded as two distinct sets of points, one set being reached if the line is approached on the upper surface from the left and on the lower surface from the right, and the other set of points being reached if the line is approached on the lower surface from the left, and on the upper from the right.

This surface is 2-sided. If, for example, we start on the upper side of the upper sheet, and move about, as we will, on the surface, we return always on the upper side of the

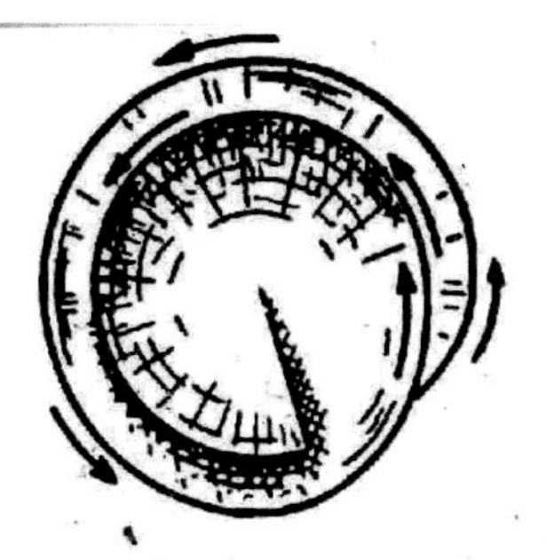

Figure 1. Two-sided surface bounded by 2-turn coil

upper sheet. This is not true for the Moebius strip.

Faraday's law then applies for this 2-sided surface of Figure 1. The integral of the normal component of B over this surface,

$$\int\int B_n dS = \Phi$$

has a meaning. Such an integral over such a surface is the only definition of the flux linkages Φ, which can be given in terms of operations which refer only to the coil in question, even though other electric circuits and sources of magnetic fields may be present.

A surface suitable for application of Faraday's law may be generated as follows. From any point fixed in space draw a straight line, radius vector, to a variable point on the bounding circuit curve. The surface generated by this moving radius vector, as the variable point traces out the bounding circuit, will be a smooth (except at the vertex) 2-sided surface to which Faraday's law may be applied.

Students of electromagnetism probably will recognize that if a layer of normally magnetized material with surface density of magnetization i, is placed uniformly along the surface of Figure 1, it will produce the same magnetic effect as a current i, flowing in the bounding circuit.

J. SLEPIAN (F'27)
(Associate Director, Westinghouse Research Laboratories, East Pittsburgh, Pa.)

M.5 Magnetic Speedometer for Aircraft

Magnetic Speedometer for Aircraft

A horizontal metallic rod, moving parallel to the earth's surface, in a direction perpendicular to itself, has induced in it, due to its motion in the earth's magnetic field, an electromotive force $V = 10^{-8}\,Hlv$ volts, where H is the vertical component of the earth's magnetic field, in gauss, l is the length of the rod in centimeters, and v is the velocity in centimeters per second. If the rod makes an angle θ with the direction of its motion, then the electromotive force is $V = 10^{-8}\,Hlv \sin\theta$. If $l = 100$, $v = 100$, corresponding to 3.6 kilometers per minute, and if $H = 0.1$, which is less than the magnitude of H for latitudes greater than 15 degrees north, and if $\sin\theta = 1$, then $V = 10^{-5}$ volts, a quantity which is readily measurable through the utilization of modern electronic instrumentation.

This effect then can be utilized for measuring the speed of an aircraft. It, of course, will give the speed of the craft relative to the earth itself, since winds or air motions generally do not affect the earth's magnetic field. This is a great advantage over other methods, using Pitot tubes or other aerodynamic instruments, since these give the craft velocity relative to the adjacent air; side drift, head, or tail winds

make such velocity relative to the air of little value [for] accurate navigation.

The electromotive force induced in the rod probably c[an] be determined best by cutting the bar at its middle a[nd] placing the potential measuring device there. We m[ay] increase the capacity to space at the ends of the rod [by] attaching large plates or spheres there. Then most of [the] induced electromotive force will appear as voltage acr[oss] the central gap in the rod.

It probably will be advantageous to rotate the rod [at] moderate speed about a vertical axis. This will make [the] voltage across the central gap alternating, and will ma[ke] easy the problem of obtaining enough insulation resista[nce] there. Also it is found much easier reliably to amplify [a] small alternating voltage than it is to amplify a direct vo[lt]age.

If the position is noted where the rotating rod has [a] maximum induced electromotive force, then the directi[on] of absolute motion of the craft may be determined, and th[us] also the amount of side-drift observed.

Access to the terminals of the central gap in the rotati[ng] rod would be obtained by slip-rings, of course.

Values of the earth's magnetic field are observed regula[rly] by the United States Department of Terrestial Magnetis[m] and maps are issued at frequent intervals. As this inv[en]tion comes into general use, we may expect these obser[va]tions to be taken more frequently, more accurately, and [at] more closely spaced points.

Is this invention sound?

Electrical Essay

Turn Ratio of Transformers

A CASE having two pairs of electrical terminals is said to contain a transformer. By steady-state 60-cycle measurements at the terminals, determine precisely the primary and secondary turn ratio. Assume that electrostatic capacity effects are negligible.

J. SLEPIAN (F '27)
(Associate director, Westinghouse Research Laboratories, East Pittsburgh, Pa.)

Answer to Previous Essays

Thevenin's Theorem. The following is the author's answer to his previously published essay (*EE, Apr '49, pp 307–08*).

The student should have short-circuited the rectifiers when taking measurements of the network impedance. The resultant resistance of 20 and 100 ohms in parallel is $16^2/_3$ ohms. The equivalent network is an electromotive force of 100 volts direct current (E_0) and a $16^2/_3$-ohm resistance (Z) in series with it. This equivalent network holds true only for an external circuit whose electromotive force, if any, does not exceed 120 volts direct current. For an external circuit with an electromotive force exceeding 120 volts direct current, the network is replaced by a 100-ohm resistance. For an external circuit with an electromotive force of negative polarity the equivalent network is 120 volts direct current and a 20-ohm resistance.

Presence of half-wave rectifiers in a circuit with variable electromotive forces introduces discontinuities in circuit constants depending on the directions and relative magnitudes of the electromotive forces. Thevenin's theorem must be applied with care within the limits of discontinuities. The essay was presented for the purpose of calling attention to this fact.

A. A. KRONEBERG (M '48)
(Southern California Edison Company, Los Angeles, Calif.)

Magnetic Speedometer for Aircraft. The following is the author's answer to his previously published essay (*EE, Apr '49, p 308*).

The invention is unsound.

When the airplane is at rest on the ground, the instrument described in the invention continues to give readings, because the rod is in the earth's electric field also. Using the language of the essay, the "electromotive force" arising from the electric field is $V=E_hl$ volts, where E_h is the horizontal component of the earth's electric field, in volts per centimeter. When in motion, the instrument presumably will read $V=10^{-8}Hlv+E_hl$. Since E_h is large and variable, the preceding equation is of little use to the pilot, for determining v.

The vertical electric gradient over the earth in fair weather is variable, but of the order of 100 volts per meter. Under a cloud, even without lightning it may be 10,000 volts per meter. Due to uneveness of the ground, and the location of the airplane relative to charged clouds, the horizontal component of the field, E_h, will have similarly large values. Hence the term in the equation for V, arising from E_h will be of the order of hundreds of volts, and will be very large compared to the term arising from H.

It might seem that some kind of shielding could be devised, which would screen the instrument from the earth's electric field, but still would permit the motion in the magnetic field to have its effect. To see that this is not the case, we must discourse on the relativity of the electromagnetic field. In this discourse we shall use centimeter-gram-second units, which should not affect its readability. E will be in electrostatic units, and H in electromagnetic units.

It must be noted that the electric and magnetic field have no absolute reality or significance, but are observed and defined only relative to some arbitrary frame chosen by the observer. The electric field usually is defined as the observed force per unit charge on a small charge at rest. But at rest relative to what? At rest relative to some arbitrary frame which the observer chooses to regard as at rest. If the observer changes his rest frame, that is, that system of material bodies which he chooses to regard as at rest, then the probe charge which he used before no longer serves to define the electric field. He must determine the force per unit charge now on another probe charge, one that is at rest relative to the new reference system, and therefore moving relative to the first probe charge. Now generally, the force per unit charge on the two probe charges both at a given point in space, but moving relatively to each other, will be different. Therefore, the observer will find two different values of the electric field, depending on whether he regards the one reference system or the other reference system as at rest.

The electric field component E_y is defined or observed by the equation

$$F_y=\rho E_y \tag{1}$$

in a reference co-ordinate system for which the probe co-ordinates (x,y,z) are unchanging, where F_y is the force component which must be applied to small probe of charge ρ to keep it at rest in the given co-ordinate system.

If a different reference rest system is used, we will need to use a different probe moving relatively to the first probe even if we retain the same charge ρ. We then will find a different F' and E', still connected by the same relation,

$$F_y'=\rho E_y' \tag{2}$$

This new E_y' is just as good in every way as the old E_y. There is no possible way of saying that one is right, and the other wrong. Each is the electric field relative to the reference frame in which its defining probe charge is at rest, and each serves to describe equally well the observed motions of charged bodies. The electric field exists and is known only relative to some arbitrarily chosen reference frame.

Similar remarks may be made about the magnetic field.

We may define it by the forces observed acting on moving probe charges. We find that we agree with experience, if we use the Lorentz formula:

$$F_y = \rho\left(E_y - \frac{1}{c} H_z v_x\right) = \rho\left(E_y - \frac{1}{c} H_z \frac{dx}{dt}\right) \quad (3)$$

where $v_x = \frac{dx}{dt}$ is the X component of velocity of the probe relative to the same reference frame which was used for defining E_y and c is a universal constant, the velocity of light. If a different reference frame is used, in which the probe charge has the co-ordinates x', y', z', then we may get a different H_z by the equation.

$$F_y' = \rho\left(E_y' - \frac{1}{c} H_z' v_x'\right) = \rho\left(E_y' - \frac{1}{c} H_z' \frac{dx'}{dt'}\right) \quad (4)$$

Suppose we know the components of E and H relative to one reference frame. Can we from these determine the components of E' and H' defined relative to another frame by the foregoing equations, if we know the motion of this second frame relative to the first frame? The answer is yes, provided we know how to calculate x', y', z', t', and F_y' from x, y, z, t, and F_y. For this last we need to know the kinematics and mechanics of the physical bodies we deal with; that is we need to know the properties of meter-sticks, clocks, and massive particles, as revealed by experience.

For relative velocities which are small compared to the velocity of light, as for example the velocity of our airplane relative to the earth, and the velocity of the particles of the rotating rod relative to the airplane, Newtonian space-time and Newtonian mechanics are well established as applicable. For reference frames in plane and earth respectively we have

$$x' = x - v_0 t;\quad y' = y;\quad z' = z;\quad t' = t;\quad F_y' = F_y;\quad m' = m;\quad \rho' = \rho \quad (5)$$

where v_0 is the constant velocity of the airplane relative to the earth, and in the X direction, and provided the reference frame in the earth is a Galilean frame, that is, one in which Newton's laws of particle dynamics holds. This last is true for our purpose, since the effect of the earth's "rotation" may be taken into account by a change in the gravitational potential, and the Coriolis forces will have an effect only over long distances or long times, as in a Foucault pendulum experiment.

In any case, for relatively moving Galilean frames with uniform relative velocity v_0, and with v_0 and any other velocities considered, small compared to that of light, equation 5 holds.

These equations give for the transformation of the velocities, applicable to any particular particle, as for instance a moving probe charge,

$$v_x' = \frac{dx'}{dt'} = \frac{dx}{dt} - v_0 = v_x - v_0;\quad v_y' = v_y;\quad v_z' = v_z \quad (6)$$

Then from the defining equations of the electric and magnetic fields, and the equality of F_y and F_y',

$$F_y' = \rho\left(E_y' - \frac{1}{c} H_z' v_x'\right) = \rho\left(E_y' - \frac{1}{c} H_z' [v_x - v_0]\right) = \rho\left(\left[E_y' + \frac{1}{c} H_z' v_0\right] - \frac{1}{c} H_z' v_x\right) \quad (7)$$

$$F_y = \rho\left(E_y - \frac{1}{c} H_z v_x\right)$$

we get

$$E_y' = E_y - \frac{1}{c} H_z v_0 \quad (8)$$

$$H_z' = H_z$$

The observer on the airplane makes experiments with charges, at rest or in motion relative to the airplane, and is able to account for what he observes by introducing electric and magnetic fields. If he likes, he may regard his airplane as at rest, and relative to axes fixed in the airplane, his experiments will reveal E' and H'. This is all his experiments can do.

On the other hand, he may prefer to use axes fixed on the earth, and relative to these axes, if he knows the value of v_0, he may determine E and H, which should agree with the electric and magnetic fields found by an observer on the earth. E, H, and E', H' are related by equations 8 and it is only through this relation that v_0 may be determined.

It should be clear now that for the observer on the airplane to determine the relative velocity of the earth from electrical experiments made only on the airplane, he must know not only H_z, but also E_y in equations 8. Since E_y referred to the earth is large and variable, the proposed speedometer for aircraft cannot be made operative.

When the velocities under consideration are not small compared to that of light, the velocities determined by meter-sticks and clocks no longer satisfy equations 6. However, we may make a modification in equations 5 changing them to the Lorentz transformation of the restricted theory of relativity, and then we obtain a modified set of equations 6 which do not disagree with the Michelson-Morley experiment on the velocity of light relative to relatively moving frames.

The Lorentz transformation replacing equation 5 does not leave the equations of Newtonian mechanics invariant. We therefore modify Newtonian particle mechanics to relativistic particle mechanics. We are then able to use those equations of equation 7 which involve only the first equality signs on each line, and arrive at transformation equations for the electric and magnetic field which are valid also for v_0 not small compared to c, the velocity of light. We get

$$E_y' = \frac{E_y - \frac{1}{c} H_z v_0}{\sqrt{1 - \frac{v_0^2}{c^2}}} \quad (9)$$

$$H_z' = \frac{H_z - \frac{1}{c} E_y v_0}{\sqrt{1 - \frac{v_0^2}{c^2}}}$$

For the airplane and earth, these equations reduce to the equations 8 since $\frac{v_0^2}{c^2}$ is negligible compared to 1, and $\frac{1}{c} E_y v_0$ is negligible compared to H_z north of latitude 15 degrees north.

J. SLEPIAN (F '27)
(Associate director, Westinghouse Research Laboratories, East Pittsburgh, Pa.)

M.6 Motionally Induced Electromotive Force–Part I

In this essay it seems that Jack, the physicist does not really under stand how current flows in conductors and thinks a single charge at the high potential side of the conductor must flow to the low potential side to produce a current. Here Jack gets into trouble with his ideas concerning the nature of current flow. Alter Ego is Dr. Slepian's evil twin who likes to ask embarrassing questions to Jack.

Electrical Essays

Motionally Induced Electromotive Force—Part I

E and B in Moving Matter

Jack, the physicist, has agreed to give a lecture to Alter Ego and his friends, explaining how electric motors work. Of course, Alter Ego is a most attentive listener. Jack has just described the slide-wire experiment of Figure 1.

Jack: "Now we have seen that the force **F** exerted by the electric and magnetic fields on a moving particle or small material body, bearing a charge q, and moving with velocity **v**, is given by the famous Lorentz formula

$$\mathbf{F}=q\left(\mathbf{E}+\frac{1}{c}[\mathbf{v}\times\mathbf{B}]\right) \tag{1}$$

Apply this relation to the electrons in the metal crossbar of Figure 1. We see that each electron, since it has to move along with the crossbar with velocity **v**, will have acting on it not only the force arising from any electric field **E**, which may be present, but also the force arising from its motion in the magnetic field, so that the total force on it will be

$$\mathbf{F}_e=e\left(\mathbf{E}+\frac{1}{c}[\mathbf{v}\times\mathbf{B}]\right) \tag{2}$$

where e is the charge on the electron.

"This force, F_e, ..."

Alter Ego, interrupting: "May I ask a question? What is the shape of the holes these electrons move in?"

Jack: "Why, Alter Ego, what do you mean?"

Alter Ego: "Well, these electrons moving around in the metal have to have some empty space to move around in, in between the atoms, or how could they move? Now, what I want to know is, what is the shape of that empty space or hole that the electron moves in. I read somewhere (*EE, Sep'50, pp 791 2*) that the force on a small charged body

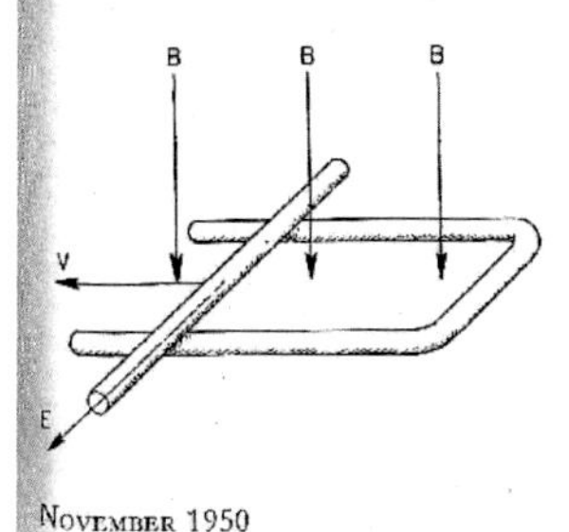

Figure 1. The slide-wire experiment

which has been placed in a hole in a body lying in an electric and magnetic field is different depending on the shape of the hole.

"For example, if the hole for the electron in the crossbar is a lozenge-shaped hole with its faces perpendicular to the magnetic field, then inside that lozenge-shaped hole the electric and magnetic fields for the directions we use will be **E** and **B**, and we would then have for the force on the electron

$$\mathbf{F}=e\left(\mathbf{E}+\frac{1}{c}[\mathbf{v}\times\mathbf{B}]\right) \tag{3}$$

corresponding to your equation 2.

"However, if the faces of the lozenge-shaped hole are perpendicular to the electric field, then the field components in the hole would be **D** and **H**, and we would have

$$\mathbf{F}=e\left(\mathbf{D}+\frac{1}{c}[\mathbf{v}\times\mathbf{H}]\right) \tag{4}$$

and if the faces of the hole are perpendicular to v, then we have

$$\mathbf{F}=e\left(\mathbf{E}+\frac{1}{c}[\mathbf{v}\times\mathbf{H}]\right) \tag{5}$$

"If the hole isn't lozenge-shaped at all, but spherical then we would have

$$\mathbf{F}=e\left(\frac{3\epsilon}{2\epsilon+1}\mathbf{E}+\frac{3}{2\mu+1}\frac{1}{c}[\mathbf{v}\times\mathbf{B}]\right) \tag{6}$$

where ϵ and μ are the dielectric constant and permeability of the crossbar material. None of these equations 4, 5, and 6 agree with your equation 1.

"So, Dr. Jack, please tell me what is the shape of the hole the electron moves in, since that will affect the force on the electron."

Jack: "You do ask the darndest questions, Alter Ego. I don't know what the shape of the holes the electrons move in is.

"Lorentz himself worked on the electron theory of matter, where all the electrical and magnetic properties of bodies were accounted for by small charged particles, the electrons and ions producing very intense local electric fields by their charge, and very intense local magnetic fields by their motion, these intense local fields, for any particular point fixed in space, also varying violently in time. Now Lorentz showed that if you take the average of all these local electric fields over a region of space embracing many of these electrons and ions, and over a time interval embracing many fluctuations in their position, then you get the **E** of Maxwell theory. But if you average all the local magnetic fields over the same space region and time interval, you get the **B** of Maxwell theory. That is why we write the force on the electron as shown in equation 2."

$$\mathbf{F}=e\left(\mathbf{E}+\frac{1}{c}[\mathbf{v}\times\mathbf{B}]\right) \tag{2}$$

(To be Continued)

Did Jack answer Alter Ego's question properly?

J. SLEPIAN (F '27)
(Westinghouse Research Laboratories, East Pittsburgh, Pa.)

Answers to Previous Essays

Motionally Induced Electromotive Force—Part I. The following is the author's answer to his previously published essay (*EE, Nov '50, pp 1025–26*).

No, I do not think Jack answered Alter Ego's question properly. Alter Ego also made some incorrect statements in his question.

Let us settle Alter Ego first. Alter Ego said that in a small lozenge-shaped hole in a body, the component of the electric field E_n^* in the hole normal to the lozenge-faces is $\mathbf{D}_n$ and the component parallel to the faces, E_s^*, is equal to $\mathbf{E}_s$ where **E** and **D** are the two Maxwellian electric fields in the body. He also asserted similarly that $H_n^*=\mathbf{B}_n$ and $H_s^*=\mathbf{H}_s$. These statements are true for a body at rest, but not for a body in motion (*EE, Nov '50, pp 1025–26*).

For a body in motion, we continue to have as defining **H** and **B**

$$\mathbf{D}_n=E_n^* \tag{1}$$

$$\mathbf{B}_n=H_n^* \tag{2}$$

These two equations ensure that Maxwell's equations concerning div **D** and div **B** will hold in all empty space, even in empty space crevices in stationary or moving bodies.

For moving bodies, however, to define **E** and **H** we have

$$\mathbf{E}_s+\frac{1}{c}[\mathbf{v}\times\mathbf{B}]_s=E_s^*+\frac{1}{c}[\mathbf{v}\times H^*]_s \tag{3}$$

$$\mathbf{H}_s-\frac{1}{c}[\mathbf{v}\times\mathbf{D}]_s=H_s^*-\frac{1}{c}[\mathbf{v}\times E^*]_s \tag{4}$$

These equations ensure that Maxwell's equations concerning curl **E** and curl **H** will hold in all empty space, including crevices in matter moving with velocity **v**. (See, for example, reference 1, and later essays in which I hope to develop this subject.) Equations 3 and 4 serve to define **E** and **H** within the moving matter.

Applying these relations to Alter Ego's lozenge shaped

holes, we have for his three orientations, assuming **B** is parallel to **H**, and **D** is parallel to **E**:

Faces perpendicular to magnetic field

$$E^* = \mathbf{E},\ H^* = \mathbf{B}, \text{ and } \mathbf{F} = e\left(\mathbf{E} + \frac{1}{c}[\mathbf{v}\times\mathbf{B}]\right) \qquad (5)$$

Faces perpendicular to electric field

$$E^* = \mathbf{D},\ H^* = \mathbf{H}, \text{ and } \mathbf{F} = e\left(\mathbf{D} + \frac{1}{c}[\mathbf{v}\times\mathbf{H}]\right) \qquad (6)$$

Faces perpendicular to velocity

$$E^* = \frac{1}{1 - v^2/c^2}\left(\mathbf{E} + \frac{v}{c}[\mathbf{B} - \mathbf{H}] - \frac{v^2}{c^2}\mathbf{D}\right)$$

$$H^* = \frac{1}{1 - v^2/c^2}\left(\mathbf{H} + \frac{v}{c}[\mathbf{D} - \mathbf{E}] - \frac{v^2}{c^2}\mathbf{B}\right)$$

$$\mathbf{F} = e\left(\mathbf{E} + \frac{1}{c}[\mathbf{v}\times\mathbf{B}]\right) \qquad (7)$$

so that Alter Ego's equation 5 was wrong.

Now turning to Jack, the physicist, I must first criticize him for referring to electron theory at all, in developing electromagnetism in moving matter as applied to the electric motor. Electric motors were running before the electron was discovered, and they have continued to run quite undisturbed as the electron changed from the charged ordinary Newtonian particle of J. J. Thomson and Lorentz to the quantum-mechanical complex probability-density wave-function, with Fermi-Dirac statistics, and negative energy states, of today. We should have and do have an electromagnetism of macroscopic bodies, with concepts defined by operations on macroscopic bodies, and sufficient for the electric motor, and independent of the current theory of the electron and the microphysics of matter.

If we know the electric and magnetic fields in empty space adjacent to bodies and in crevices in bodies, and if Maxwell's equations hold in this empty space, then equations 1, 2, 3, and 4 will define vectors **D**, **B**, **E**, and **H** at points inside matter, and Maxwell's equations will hold for these vectors with ρ defined by $4\pi\rho = \text{div } \mathbf{D}$, and with defined by $\frac{4\pi}{c}\left(\mathbf{i} + \frac{1}{4\pi}\frac{\partial \mathbf{D}}{\partial t}\right) = \text{curl } \mathbf{H}$.

We then look for constitutive relations between these field quantities which depend only on the nature of the material involved, but now, along with pressure, temperature, and so forth, which are parameters defining the nature of the material, we must include the velocity **v**. Thus we may find that experiment verifies the relation

$$\mathbf{i} = \sigma\left(\mathbf{E} + \frac{1}{c}[\mathbf{v}\times\mathbf{B}]\right) \qquad (8)$$

for metals for which at rest, $\mathbf{i} = \sigma\mathbf{E}$. Then equation 8 is a constitutive equation for such metals in motion. But enough of this for now. There will be more another time.

Let us consider Jack the physicist's argument based on the Lorentz theory of the electrical structure of matter. Let us accept Lorentz's postulate that matter is made of charged particles, electrons and ions, which behave like charged macroscopic small bodies, having definite position in space and time, producing electric and magnetic fields, and being acted upon by nonelectrical or mechanical forces as well as the forces of the electric and magnetic field, even though it is now well known that the individual electrons and ions do not behave at all like macroscopic small charged bodies.

Then under these conditions, Lorentz showed that for a medium, at rest, the average of the microscopic electric fields over an extended region would be the Maxwell field **E**, and the average of the microscopic magnetic fields would be Maxwell's **B**. But is this true also for a medium in motion? The moving electrons and ions will produce different microscopic fields than when they are at rest, and it is not at all clear these fields will continue to average to Maxwell's **E** and **B**.

More serious than this, however, is the question as to whether the electron is to be regarded as in the average fields **E** and **B**. The electrons are not scattered through the metal completely at random. They will tend to favor some interatomic regions and avoid others. Thus the average field that an electron finds itself in is not the same as the average field throughout a whole region, and therefore the electron cannot be said to be in either Maxwell's **E** field or Maxwell's **B** field.

The reasoning of Jack, the physicist, was not sound. Alter Ego's question, "What is the shape of the holes the electrons move in?" has not been answered.

REFERENCE

1. **Electrodynamik, Sommerfeld. Dieterichsche Verlagsbuchhandlung,** 1948, page 288.

J. SLEPIAN (F '27)
(Westinghouse Research Laboratories, East Pittsburgh, Pa.)

Switching Gravity On and Off. The following is the author's answer to his previously published essay (*EE, Nov '50, p 1026*).

The scientist was the famous physicist, Dr. Max Steenbeck, of the Siemens Schuckert Company, in Berlin.

Dr. Steenbeck placed the arc, together with a movie camera, inside a box, and, with proper flexible leads for power and oscillographic recording, tossed the box upward to a height of about ten feet, letting it fall freely back again. With respect to co-ordinate axes moving with the box, during the free flight, there was no gravitational field. Thus within the box, gravity was turned off at the beginning of the flight, and turned on again at the end.

Dr. Steenbeck describes the interesting results he obtained in a paper.[1] In this paper, Dr. Steenbeck refers to a similar experiment made with respect to a candle by H. Lorenz of Bonn, Germany.

Professor Lorenz, in his paper,[2] states that he saw Professor W. Gerlach of the University of Munich perform a similar experiment in a lecture demonstration in 1932.

REFERENCES

1. **Investigation of Electric Arc in Gravity-Free Space, Max Steenbeck.** *Physikalische Zeitschrift* (Leipzig, Germany), volume 38, 1937, pages 1019–21.

2. **Demonstration of "Inertia Force" on a Freely Falling Candle Flame, H. Lorenz.** *Physikalische Zeitschrift* (Leipzig, Germany), volume 35, 1934, page 529.

J. SLEPIAN (F '27)
(Westinghouse Research Laboratories, East Pittsburgh, Pa.)

M.7 Motionally Induced Electromotive Force–Part II

In this essay Jack, the physicist is trying to explain the principles of electric motors and generators by introducing the Hall effect. He gets into great difficulty here.

Electrical Essays

Motionally Induced Electromotive Force—Part II

The Hall Effect

Jack, the physicist, is continuing his lecture to Alter Ego and his friends explaining the principles of electric motors and generators (see Part I, *EE, Nov '50, pp 1025–26*).

Jack: "As I was saying before Alter Ego interrupted me, the electrons in the cross bar of Figure 1 have to move with its velocity **v** in the direction perpendicular to the magnetic field, and therefore will have acting on them on this account a force

$$F = e\frac{1}{c}[\mathbf{v}\mathrm{x}\mathbf{B}] \qquad (1)$$

"Since this force is proportional to the charge e, it produces the same effect as if there were an electric field acting on it:

$$\mathbf{E}_{mot} = \frac{1}{c}[\mathbf{v}\mathrm{x}\mathbf{B}] \qquad (2)$$

"We'll call $\mathbf{E}_{mot}$ of equation 2 the motional field, or motionally induced electric field. If we integrate it from one point to another, we'll call the result the motionally induced electromotive force, although if **v** is not constant throughout the material, this integral may depend on the path of integration.

"In a moving metal, then, since not only the regular electric field **E** but also $\mathbf{E}_{mot}$ will be acting on the electrons, the current density will not be given by Ohm's law

$$\mathbf{i} = \sigma\mathbf{E} \qquad (3)$$

where σ is the conductivity of the metal, but we will have instead

$$\mathbf{i} = \sigma\left(\mathbf{E} + \frac{1}{c}[\mathbf{v}\mathrm{x}\mathbf{B}]\right) = \sigma(\mathbf{E} + \mathbf{E}_{mot}) \qquad (4)$$

"If we make the current zero, as by opening the circuit of Figure 1, then we must have $\mathbf{E} + \mathbf{E}_{mot} = 0$, and $\mathbf{E} = -\mathbf{E}_{mot} = -\frac{1}{c}[\mathbf{v}\mathrm{x}\mathbf{B}]$. To get the open circuit voltage then, since $\mathbf{E} = -$gradient V where V is the electrostatic potential, we integrate through the wires of Figure 1 from the one open end to the other, and get

$$V_o = \int -\mathbf{E}\cdot d\mathbf{s} = \frac{1}{c}vBl \qquad (5)$$

where l is the length of the slide wire between the two sliding contact points.

"If I close the circuit, the total current I will be given by

$$RI = V - V_o = V - \frac{1}{c}vBl \qquad (6)$$

where R is the resistance of the circuit.

"Multiplying equation 6 by I, we get

$$VI = RI^2 + V_oI = RI^2 + \frac{1}{c}vIBl \qquad (7)$$

where VI is the electrical power input, RI^2 is the Joulean heat developed, and $V_oI = \frac{1}{c}vIBl$ is the mechanical power, since $\frac{1}{c}IB$ is the force which acts per unit length on a conductor carrying a current I, and which lies in a perpendicular magnetic field **B**. But I see that Alter Ego has been trying to ask a question. What is it, Alter Ego?"

Alter Ego: "This **v** in your equations is the velocity of the electrons in the magnetic field, which you produced by moving the slide bar, but there are other ways of giving electrons velocity. For example, in a piece of metal, at rest, if I send current in it, I'll have

$$\mathbf{i} = ne\mathbf{v} \text{ or } \mathbf{v} = \mathbf{i}/ne \tag{8}$$

where n is the density of electrons in the metal.

"Now, if I put this metal carrying current in a magnetic field, even though the metal is at rest, the electrons are moving, and I'll get a motional field

$$\mathbf{E}_{mot} = \frac{1}{c}[\mathbf{v}\mathrm{x}\mathbf{B}] = \frac{1}{cne}[\mathbf{i}\mathrm{x}\mathbf{B}] \tag{9}$$

Is that right?"

Jack: "Well, we do not usually call it a motional field, but what you have done is to independently discover the Hall Effect. Not only that but you have even given a formula for the Hall Effect coefficient, $1/cne$, which is just about right. That is wonderful, Alter Ego!"

Alter Ego: "I don't know how big n is, but at least I know it is positive, and c is positive, and e is negative, so I can be very sure that the Hall Effect coefficient is always negative, isn't that right, Dr. Jack?"

Jack: "Well no, some metals, zinc, cadmium, and lead, for example, have Hall Effect coefficients of the opposite sign."

Alter Ego: "Gosh! Then the electrons in these metals have a positive charge?"

Jack: "No. You see, the electrons in the metal are not quite as simple as I made out. They are really wave functions in the periodic lattice potential and because they are degenerate, they satisfy Fermi-Dirac statistics, and lie in the bottom portion of a band of allowed energy levels. Some of the electrons get pushed into the upper energy levels and leave holes in the assembly of electrons in the lower levels, and these holes act like electrons with positive charge."

Alter Ego: "Double gosh! Then you can have a force like $\frac{1}{c}[\mathbf{v}\mathrm{x}\mathbf{B}]$ acting on a hole?"

Jack, weakly: "Yes."

Alter Ego: "Well, anyway, since you told me the Lorentz force formula is all right, I can be sure that the Hall Effect is proportional to **B**, isn't that right?"

Jack: "Well, no. The Hall Effect in the magnetic metals is not proportional to either **B** or **H**. The Hall Effect is quite complicated in these metals."

Alter Ego: "Then, when calculating the motional electromotive force of a conducter moving in a magnetic field, like in the slide-wire experiment, you say that the electrons are small bodies with charge e, moving in the Maxwell field **B**, subject to the Lorentz force, $\mathbf{F} = e\frac{1}{c}[\mathbf{v}\mathrm{x}\mathbf{B}]$, thus giving the motionally induced electric field $\mathbf{E}_{mot} = \frac{1}{c}[\mathbf{v}\mathrm{x}\mathbf{B}]$, which is the same for all metals; but when I move the electrons in the stationary metal to calculate the Alter Ego Effect, also known as the Hall Effect, then you say the electrons become waves, or holes, their charge is no longer the negative number e, they no longer move in Maxwell's **B** or even **H**, so they certainly can't have acting on them the Lorentz force, $e\frac{1}{c}[\mathbf{v}\mathrm{x}\mathbf{B}]$. I am confused."

Jack: "I am a bit confused, too. I knew that the simplified electron theory I gave you was wrong, but at least you would understand it. Instead I should have given you the true theory, which you could not possibly understand."

(*To be continued.*)

There are several questions to be asked: 1. Should Jack have appealed to electron theory at all? 2. Did Jack present the simple Lorentz picture correctly? 3. Is there not some more general macroscopic principle or theory that Jack might have appealed to rather than the electron theory of the microstructure of matter?

J. SLEPIAN (F '27)
(Westinghouse Research Laboratories, East Pittsburgh, Pa.)

Answers to Previous Essays

Motionally Induced Electromotive Force—Part II. The following is the author's answer to his previously published essay (*EE, Dec '50, pp 1086–87*).

> "O, what a tangled web we weave,
> When first we practice to deceive."

Question 1. Poor Jack! He should not have appealed to electron theory at all. He knew that the simple Lorentz electron theory picture of matter is not true, but it had the great merit of being understandable to his students, Alter Ego, and his friends. How many teachers face this dilemma! To give a false theory, which the students will say they understand, or a more sophisticated and truer statement which will be incomprehensible to the immature student?

Questions 2 and 3. No, Jack did not present the simple Lorentz picture correctly, and yes, there is a more general macroscopic principle that Jack might have appealed to, and then he could have left the electrons out. In fact, Jack tacitly used this general principle in his faulty treatment of the Lorentz theory, although of course that is not what made his treatment faulty.

Now what Jack is after, of course, is the constitutive relation for a metal in motion, assuming that for a metal at rest it is

$$\mathbf{i}=\sigma\mathbf{E} \qquad (1)$$

Now, Jack did assume in his discussion that in some sense the moving bar had exactly the same properties as the stationary bar, and that relative to the electrons inside it, or relative to any other objects moving along with it, the bar would behave the same as the bar at rest, relative to the electrons within it, or relative to other objects at rest with it. But this is the principle of relativity, and Jack did not need to go farther to derive the constitutive equation for the moving bar. Knowing the constitutive equation for the bar at rest, say Ohm's law, or equation 1, then relativity alone is sufficient to determine the constitutive equation for the bar in motion, and further inquiry into the details of microscopic electron theory is unnecessary, and for the desired purpose, irrelevant.

If we include among the objects moving along with the moving bar an observer, B, then the principle of relativity just described would assert that to the observer B the bar would appear, through such experiments as B could make, to have the same constitutive equation as that for the bar at rest, as determined by an observer A, also at rest with the bar.

Therefore, if B sees the field $\mathbf{E}'$, and the current $\mathbf{i}'$, B will find the constitutive equation

$$\mathbf{i}'=\sigma\mathbf{E}' \qquad (2)$$

But the fields $\mathbf{E}'$, and so forth, and the charge and current densities ρ', $\mathbf{i}'$, which B sees will not be the same as the fields $\mathbf{E}$, and so forth, and the charge and current densities, ρ, $\mathbf{i}$, which A sees, since B and A will differ as to what is the velocity of the charged probes which they use for observing the fields. If A and B use relativistic particle mechanics, the field $\mathbf{E}'$ will be related to the fields $\mathbf{E}$ and $\mathbf{B}$ by the equations

$$\mathbf{E}_x'=\mathbf{E}_x$$

$$\mathbf{E}_y'=\left(\mathbf{E}_y-\frac{1}{c}v\mathbf{B}_z\right)\left(1-\frac{v^2}{c^2}\right)^{-1/2}$$

$$\mathbf{E}_z'=\left(\mathbf{E}+\frac{1}{c}v\mathbf{B}_y\right)\left(1-\frac{v^2}{c^2}\right)^{-1/2} \qquad (3)$$

where the velocity $\mathbf{v}$ is entirely in the x direction, and is of magnitude v.[1] If we neglect v^2/c^2 compared to 1, then equation 3 reduces to

$$\mathbf{E}'=\mathbf{E}+\frac{1}{c}[\mathbf{v}\times\mathbf{B}] \qquad (4)$$

which is also what we would get if we used Newtonian particle mechanics.

Similarly, if we used relativistic particle mechanics and neglect v^2/c^2 compared to 1, we get

$$\mathbf{B}'=\mathbf{B}-\frac{1}{c}[\mathbf{v}\times\mathbf{E}] \qquad (5)$$

If we had used Newtonian particle mechanics, with the definition we have given for $\mathbf{B}$, we would have had

$$\mathbf{B}'=\mathbf{B} \qquad (5A)$$

However, equation 5 with the other related primed vectors satisfies Maxwell's equations except for terms involving v^2/c^2, whereas equation 5A would fail to satisfy Maxwell's equations by terms involving v/c.

Similarly for the other field quantities, neglecting v^2/c^2 compared to 1, we have

$$\mathbf{H}'=\mathbf{H}-\frac{1}{c}[\mathbf{v}\times\mathbf{D}] \qquad (6)$$

$$\mathbf{D}'=\mathbf{D}+\frac{1}{c}[\mathbf{v}\times\mathbf{H}] \qquad (7)$$

$$\mathbf{i}'=\mathbf{i}-\frac{1}{c}\rho\mathbf{v} \qquad (8)$$

$$\rho'=\rho+\frac{1}{c}\mathbf{v}\cdot\mathbf{i} \qquad (9)$$

Equations 8 and 9 may seem strange, suggesting that relatively moving observers will disagree as to the charge and current densities in a given body, but they are true, (except for higher powers of v/c) and to be expected, as will be developed in later essays.

Now if observer B finds Ohm's Law or equation 2 holding for the metal body at rest relative to him, then observer A will find for that same body moving with velocity $\mathbf{v}$ relative to him

$$\mathbf{i}-\frac{1}{c}\rho\mathbf{v}=\sigma\left(\mathbf{E}+\frac{1}{c}[\mathbf{v}\times\mathbf{B}]\right) \qquad (10)$$

obtained by substituting equations 4 and 8 in equation 2.

Observer B may also find that the charge density ρ' in a homogeneous metallic body at rest relative to him is always zero. If we set $\rho'=0$ in equation 9, then, except for higher powers of v/c, equation 10 will reduce to

$$\mathbf{i}=\sigma\left(\mathbf{E}+\frac{1}{c}[\mathbf{v}\times\mathbf{B}]\right) \qquad (11)$$

nus, we get equation 11 which is basic for the usual treatment of voltages induced in moving wire coils as given in electrical engineering texts, but we see that it is limited to coils of material which at rest satisfy Ohm's Law, equation 2, have zero charge density, and move so slowly that v^2/c^2 is negligibly small compared to 1.

REFERENCE

1. Electromagnetic Theory (book), Stratton. McGraw-Hill Book Company, Inc., New York, N. Y., 1941, page 79.

J. SLEPIAN (F '27)
(Westinghouse Research Laboratories, East Pittsburgh, Pa.)

M.8 Motionally Induced Electromotive Force–Part III

Here Jack realizes he shouldn't have used electron theory to explain motors and generators and accepts an electric field intensity is induced in any body that moves in a magnetic field. However, his troubles with Alter Ego are not over.

Electrical Essays

Motionally Induced Electromotive Force—Part III

Motional Electromotive Force in a Vacuum

Jack, the physicist, is continuing his lectures to Alter Ego and his friends on the basic principles underlying electric motors and generators.

Jack: "I am sorry that this subject of the Hall Effect came up (*EE, Dec '50, pp 1086–87*). I guess now that I'll just have to tell you that we may regard it as well established by experiment and experience that there is a motional electric field $\mathbf{E}_{mot}$ induced in any body which moves in a magnetic field $\mathbf{B}$, and $\mathbf{E}_{mot}$ is given by the equation

$$\mathbf{E}_{mot}=\frac{1}{c}[\mathbf{v}\times\mathbf{B}] \tag{1}$$

In this equation, $\mathbf{v}$ is the ordinary everyday velocity of the material of the body and has no reference to the motion of electrons or any other theoretically existing microscopic parts of the body. $\mathbf{E}_{mot}$ acts on the body just like any other electric field such as $\mathbf{E}_s$, the electrostatic field produced by charges, and $\mathbf{E}_{ind}$, the electric field produced by magnetic fluxes varying in time, or in other words, by transformer action. The current density in the body, if it satisfies Ohm's Law, will be given by

$$\mathbf{i}=\sigma(\mathbf{E}_s+\mathbf{E}_{ind}+\mathbf{E}_{mot}) \tag{2}$$

"I now come to an interesting relation. Integrate $\mathbf{E}_s+\mathbf{E}_{ind}+\mathbf{E}_{mot}$ around a closed path which goes through the material bodies and which is being carried along with the matter in the bodies. For $\mathbf{E}_{ind}$ we have of course

$$\int \mathbf{E}_{ind}\cdot d\mathbf{s}=-\int\int\frac{1}{c}\frac{\partial\mathbf{B}}{\partial t}\cdot d\mathbf{S} \tag{3}$$

where on the right we have the rate of change of the normal component of $\mathbf{B}$, integrated over any surface which is momentarily bounded by the moving curve of integration.

"For $\mathbf{E}_{mot}$ we have from equation 1

$$\int \mathbf{E}_{mot}\cdot d\mathbf{S}=\int\frac{1}{c}[\mathbf{v}\times\mathbf{B}]\cdot d\mathbf{s}=-\int\frac{1}{c}\mathbf{B}\cdot[\mathbf{v}\times d\mathbf{s}] \tag{4}$$

But $[\mathbf{v}\times d\mathbf{s}]$ is the rate of increase of the area bounded by the closed curve due to the motion of the element $d\mathbf{s}$. Therefore, $\int\mathbf{B}\cdot[\mathbf{v}\times d\mathbf{s}]$ of equation 4 is that rate of increase of $\int\int\mathbf{B}\cdot d\mathbf{S}$ taken over the bounded surface, which arises from the motion of the bounding curve itself. Now remembering that $\int\mathbf{E}_s\cdot d\mathbf{s}=0$, we have, combining with equations 3 and 4 for the total field, $\mathbf{E}_{tot}$, acting on the material of the moving bodies, and producing effects like current flow, or dielectric polarization

$$\int\mathbf{E}_{tot}\cdot d\mathbf{s}=\int(\mathbf{E}_s+\mathbf{E}_{ind}+\mathbf{E}_{mot})\cdot d\mathbf{s}=-\int\int\frac{1}{c}\frac{\partial\mathbf{B}}{\partial t}\cdot d\mathbf{S}-\int\frac{1}{c}[\mathbf{v}\times\mathbf{B}]\cdot d\mathbf{s} \tag{5}$$

But the right-hand side of equation 5 reduces to $-1/c$ times the total rate of increase in $\int\int\mathbf{B}\cdot d\mathbf{S}$, that due to rate of change of $\mathbf{B}$ itself, that is to $\partial\mathbf{B}/\partial t$, and that due to the motion of the boundary curve. Thus we have

$$\int\mathbf{E}_{tot}\cdot d\mathbf{s}=-\frac{1}{c}\frac{d}{dt}\int\int\mathbf{B}\cdot d\mathbf{S} \tag{6}$$

where the plain d/dt means the total rate of change of the flux linkage of the circuit, $\int\int\mathbf{B}\cdot d\mathbf{S}$, including the effect of the motion of the bounding curve itself.

"This is Faraday's Law of Induction for moving bodies which says that the total electromotive force around a circuit, or $\int\mathbf{E}_{tot}\cdot d\mathbf{s}$, is equal to $-1/c$ times the rate of change of the flux linkage regardless of whether that electromotive force is due to motion or to transformer action. Faraday found that he got the same currents induced in a coil

$$I=\frac{1}{R}\int\mathbf{E}_{tot}\cdot d\mathbf{s} \tag{7}$$

regardless of whether he moved a magnet relative to the coil, or the coil relative to the magnet.

"But now I am not at all surprised to see that Alter Ego has been trying to ask some questions."

Alter Ego: "You say now that the motional electric field $\mathbf{E}_{mot}$ is the same for all bodies regardless of their internal constitution and is given by equation 1. If I use a glass rod as a slide bar in the slide wire experiment, will I get this motionally induced field in it which will induce an electric induction $\mathbf{D}$ in it like a purely electrostatic field of the same magnitude?"

Jack: "Well, if I hadn't learned my lesson from our discussion of the Hall Effect I would have said no, because I know that in the case of the electrostatic field any one dipole is not in the field which is impressed upon the glass, since the neighboring dipoles cause the field to be different at the one dipole. Actually the field at this dipole is $\mathbf{E}_s+\frac{4\pi}{3}\mathbf{P}=\frac{\epsilon+2}{3}\mathbf{E}_s$, where $\mathbf{P}$ is the polarization density of the material. This is known as the Mossotti effect. However, for the motional field, $(1/c)[\mathbf{v}\times\mathbf{B}]$, $\mathbf{B}$ will not be shielded or modified by the adjacent dipoles so I would have supposed that the motional field, $\mathbf{E}_{mot}$, would produce a different effect than a purely electrostatic field $\mathbf{E}_s$. However, I'll stick to my general principle and say that the motional field $\mathbf{E}_{mot}=(1/c)[\mathbf{v}\times\mathbf{B}]$ will produce exactly the same effect as an electrostatic field $\mathbf{E}_s$."

Alter Ego: "Then since an electrostatic field $\mathbf{E}_s$ acting on the stationary glass rod would produce an induction $\mathbf{D}$ given by

$$\mathbf{D}=\epsilon\mathbf{E}_s \tag{8}$$

for the glass rod moving in the magnetic field we should have

$$\mathbf{D}=\epsilon\frac{1}{c}[\mathbf{v}\times\mathbf{B}] \tag{9}$$

shouldn't we?"

Jack: "Yes, but of course there will be end effects which will produce an electrostatic field $\mathbf{E}_s$ which will partly oppose $\mathbf{E}_{mot}$ so that we will have

$$\mathbf{D}=\epsilon\left(\mathbf{E}_s+\frac{1}{c}[\mathbf{v}\times\mathbf{B}]\right) \quad (10)$$

Alter Ego: "I can get rid of the end effects by using a dielectric ring which can stretch, like rubber, so the motion of the material will be radial as I change the radius of the ring. I'll still have a motional field $\mathbf{E}_{mot}=(1/c)[\mathbf{v}\times\mathbf{B}]$ acting circumferentially in the ring?"

Jack: "Why, yes."

Alter Ego: "Suppose this ring is an insulating tube, like an inner tube, with air in it. As the ring stretches, the air in it will also move with a radial velocity $\mathbf{v}$, so in the air, too, according to your equation 1 there will be a circumferential motional field, $\mathbf{E}_s=(1/c)[\mathbf{v}\times\mathbf{B}]$."

Jack: "I must say that would be true. My equation 1 is universal."

Alter Ego: "Now suppose the tube material can stretch but is sufficiently rigid so that I can pump the air out. Since your equation 1 does not have in it the pressure or density of the material which is moving, $\mathbf{E}_{mot}$ will not be affected by my pumping. When I get down to a vacuum then, I'll still have the full $\mathbf{E}_{mot}=(1/c)[\mathbf{v}\times\mathbf{B}]$ in it?"

Jack: "Well, somehow I do not like your conclusion that there will be a motional field in a vacuum. I am not sure that you can say that the vacuum has a velocity $\mathbf{v}$, although I must admit it moves along with the tube. You have me stumped, Alter Ego. I do not know whether there will or will not be a motional field $\mathbf{E}_{mot}$ in the vacuum. I am sure that equation 1 holds for all honest-to-goodness matter. But, how low the density has to be for matter to stop being honest-to-goodness matter for which equation 1 always holds, I do not know."

Well, Jack is to be commended for his honesty and frankness. Can the reader clear up the difficulty in this discussion? Can there be a motionally induced electric field in a vacuum?

J. SLEPIAN (F '27)
(Westinghouse Research Laboratories, East Pittsburgh, Pa.)

68

M.9 Motionally Induced Electromotive Force–Part IV

Electrical Essays

Motionally Induced Electric Fields—Part IV

Motional Electromotive Force in Iron, Commutatorless D-C Motor, and Space Ship

Jack, the physicist, is continuing his lectures to Alter Ego and his friends on the basic principles underlying dynamo-electric machines.

Jack: "We now have the universal equation for the motional electric field, $\mathbf{E}_{mot}$, induced in any honest-to-goodness body moving with a velocity $\mathbf{v}$ in a magnetic field $\mathbf{B}$, namely

$$\mathbf{E}_{mot} = \frac{1}{c}[\mathbf{v} \times \mathbf{B}] \qquad (1)$$

This universal equation is consistent with the principle of energy, for if $\mathbf{i}$ is the current density in the body, we will have

$$P_e = \mathbf{E}_{mot} \cdot \mathbf{i} = [\mathbf{v} \times \mathbf{B}] \cdot \mathbf{i} = \mathbf{v} \cdot [\mathbf{B} \times \mathbf{i}] \qquad (2)$$

$P_e = c\mathbf{E}_{mot} \cdot \mathbf{i}$ is the electric power supplied per unit volume by the moving body external circuit.

"We find the mechanical power by means of the universal equation

$$\mathbf{F}_m = [\mathbf{i} \times \mathbf{B}] = -[\mathbf{B} \times \mathbf{i}] \qquad (3)$$

which gives the mechanical force, per unit volume $\mathbf{F}_m$, which the magnetic field $\mathbf{B}$ exerts upon any body carrying a current density $\mathbf{i}$. The mechanical power, per unit volume, which the moving body delivers to other bodies is then

$$P_m = \mathbf{v} \cdot \mathbf{F}_m = -\mathbf{v} \cdot [\mathbf{B} \times \mathbf{i}] \qquad (4)$$

"Combining equations 2 and 4 we have

$$P_e + P_m = 0 \qquad (5)$$

That is, if the moving body is delivering electric power, as in a generator, then equal mechanical power must be supplied to it, and if the body is receiving electric power by $\mathbf{i}$ flowing against $\mathbf{E}_{mot}$, as in a motor, then it will deliver an equal mechanical power. Universal equations 1 and 3 are entirely in order as far as the principle of energy is concerned.

"But now, Alter Ego has something to say."

Alter Ego: "I am very glad to learn of the universal validity of equations 1 and 3 because based on them I have made two very important inventions; namely, a commutatorless d-c motor or generator, and a d-c driven space ship. I won't go into the details of these inventions, but will limit myself to describing their basic principles.

"Go back to the slide-wire experiment, Figure 1, and consider what will happen if the slide bar is a cylinder of very high permeability iron. The lines of force of the otherwise uniform magnetic field $\mathbf{B}$ will converge as is shown in Figure 2, and $\mathbf{B}$, within the circular section iron bar, will have almost twice the strength which it would have in a similarly placed copper bar. Therefore, according to your universal equation 1 which you say is absolutely right, the motional field $\mathbf{E}_{mot}$ in the iron bar will be twice what you would get for a copper bar.

"Now for my invention. Make a coil like I show in Figure 3, with two parallel sides, but make one of these

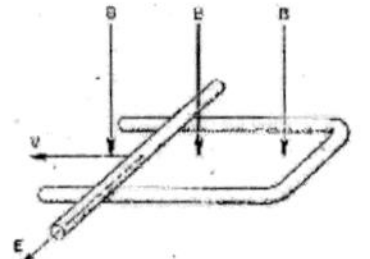

Figure 1 (above). The slide-wire experiment

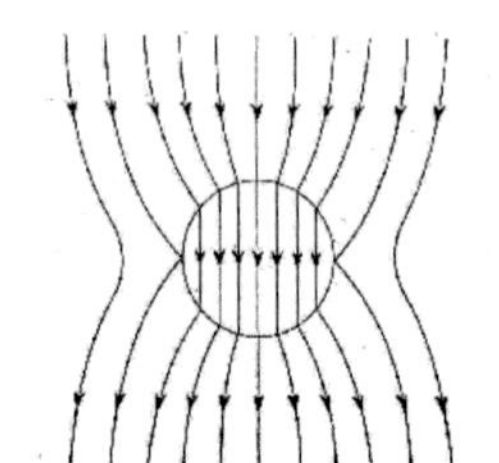

Figure 2 (right). Converging field in iron cylinder

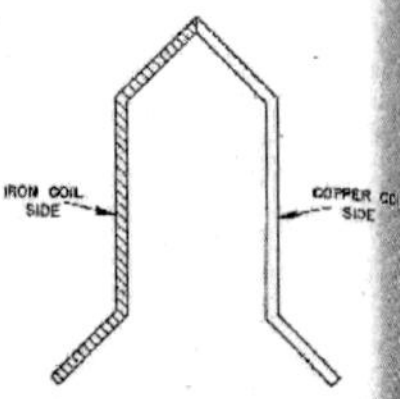

Figure 3. Iron-copper coil

parallel sides of iron and the other of copper. Now translate the coil in the magnetic field. There will be motional fields induced in the coil sides. If the coil had been all copper, then the motional field $\mathbf{E}_{mot}$ in the one coil side would oppose and just cancel the motional field $\mathbf{E}_{mot}$ in the other coil side, so far as the production of a total electromotive force around the coil is concerned. However, with one coil side iron, the motional field $\mathbf{E}_{mot}$ in it will be twice the motional field in the other side, copper coil, and these motional fields will not cancel round the coil, but the coil will give a net, not zero, electromotive force as it is moved in the uniform magnetic field. It is easy to see how we could mount a lot of these coils in series on a rotating member with slip rings, in a uniform magnetic field, and get a beautiful constant d-c voltage. There's your commutatorless d-c generator.

"Now, consider what happens if I send current through my (patent-pending) iron-copper coil while it is in a uniform magnetic field. If the coil had been all copper, then, according to your universal and absolutely right equation 3, the forces exerted by the magnetic field on the coil sides would be equal and opposite and would exactly cancel. However, in my iron-copper coil, the field **B** in the iron side is twice what it is in the copper side, and therefore, according to your universal equation 3, the force on the iron side will be twice the oppositely directed force on the copper side, and I will get a net force tending to move the whole coil. If I arrange a lot of these coils around on a rotor as in my generator, I'll get a commutatorless d-c motor. However, if I lay the coils out in a plane, I'll get a space ship, since the reaction on the field-producing member will be the same as if the coils were all copper, namely, zero. Aren't those wonderful inventions, Dr. Jack?"

Jack: "I don't like to throw cold water on your enthusiasm, Alter Ego, but I doubt that your inventions will work. You will remember that in my last lecture (*EE, Jan '51, pp 67–68*), I showed that the sum of all the electric fields, the electrostatic, that induced by transformer action, and the motional, when integrated round any closed material circuit, was equal to $1/c$ times the rate of change of the integral of **B** over any surface enclosed by the circuit, or in other words, to $1/c$ times the rate of change of the total flux linked by the circuit. Now, as your coil moves in the uniform magnetic field, the amount of flux linked by it does not change. Therefore, the total current-producing electromotive force which is around your coil stays zero.

"Of course, you should have obtained the same result using the motional equation 1, but you did not. Some people think that the **v** in equation 1 stands for the motion of the moving body, not relative to space or some suitable material frame of reference but relative to the magnetic field itself, and that where the field lines bunch up as in your iron bar the bar carries them along somewhat, so that the lines of force slide or cut through the bar with only half that velocity with which they are cut by the copper bar so that the motional field is the same in the iron as the copper. Maybe that is the way out...."

I interrupt Jack at this point, because I am already behind in my reply to the preceding essay and I must catch up.

Will Jack be able to explain away the contradiction along the line he has indicated? Watch for the next exciting episode in the next installment.

J. SLEPIAN (F [illegible])

(Westinghouse Research Laboratories, East Pittsburgh, Pa.)

Three Impedances

Three impedances are so proportioned that when connected in star and energized from a 60-cycle system

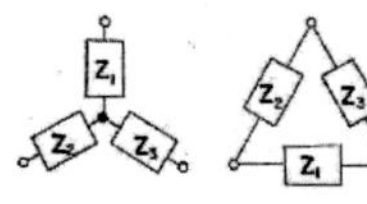

Figure 1. Star and delta connections of three impedances

they form a network that is equivalent to a network formed by the same three impedances connected in delta. Is this possible?

A. A. KRONEBERG (F [illegible])

(Southern California Edison Company, Los Angeles, Calif.)

Answers to Previous Essays

Correction to Motionally Induced Electromotive Force. The following is the author's correction to the Part II essay and the answer to Part I (*EE, Dec '50, pp 1086–9*).

The units used in these essays on "Motionally Induced

Answers to Previous Essays

Motionally Induced Electric Fields—Part IV. The following is the author's answer to his previously published essay (*EE, Feb '51, pp 159–60*).

As the author has observed so many times in the past in these various essays, relative to any particular suitable frame of reference, there is but one electric field **E**, and also but one **D**, one **H**, and one **B**, and these are the quantities which enter into Maxwell's field equations. These vectors may be defined in empty space through local observation of the forces on charged probes, placed in the empty space, by equation 3,[1] and may be defined within bodies stationary or moving through observation of the forces on charged probes within crevices in the bodies by equations 2 to 4.[2] The charge and current densities, ρ and **i**, are then defined from the quantities **D** and **H** by equations 3 and 10.[3]

Maxwell's field equations are not enough to fix or determine in a mathematical sense the various field quantities which we have defined operationally through observations on charged material probes in empty space. To get a mathematically complete system we adjoin the constitutive equations of the various materials in question, which are determined by experiment or otherwise.

If we know the constitutive equations for the matter of a body at rest, then the principle of relativity enables us to determine the constitutive equations for the body in a state of uniform motion. We merely make the transformation of the field quantities called for by relativity.[4] In this way, we obtained the constitutive equations for the glass of a glass slide bar.

$$\mathbf{D}=\epsilon\mathbf{E}+(\epsilon-1)\frac{1}{c}[\mathbf{v}\times\mathbf{B}] \tag{1}$$

$$\mathbf{B}=\mathbf{H}-\frac{\epsilon-1}{\epsilon}\frac{1}{c}[\mathbf{v}\times\mathbf{D}] \tag{2}$$

$$\mathbf{i}=\frac{1}{c}\mathbf{v}\rho \tag{3}$$

To deal with the iron slide bar, we similarly determine the constitutive equations for moving iron from the known equations for iron at rest, that is

$$\mathbf{i}'=\sigma\mathbf{E}' \tag{4}$$

$$\rho'=0 \tag{5}$$

$$\mathbf{D}'=0 \tag{6}$$

$$\mathbf{B}'=f(\mathbf{H}') \tag{7}$$

Making the substitutions of equations 1,[4] we get these equations:

$$\mathbf{i}-\frac{1}{c}\mathbf{v}\rho=\sigma\left(\mathbf{E}+\frac{1}{c}[\mathbf{v}\times\mathbf{B}]\right) \tag{8}$$

$$\rho+\frac{1}{c}\mathbf{v}\cdot\mathbf{i}=0 \tag{9}$$

$$\mathbf{D}+\frac{1}{c}[\mathbf{v}\times\mathbf{H}]=0 \tag{10}$$

$$\mathbf{B}-\frac{1}{c}[\mathbf{v}\times\mathbf{E}]=f\left(\mathbf{H}-\frac{1}{c}[\mathbf{v}\times\mathbf{D}]\right) \tag{11}$$

Neglecting higher powers of v/c than the first, we get that one of the constitutive equations of which we shall make use in this essay, namely

$$\mathbf{i}=\sigma\left(\mathbf{E}+\frac{1}{c}[\mathbf{v}\times\mathbf{B}]\right) \tag{12}$$

Maxwell's field equations plus the constitutive equations just mentioned are sufficient for solving the two slide-bar problems which Alter Ego has presented to us. For this purpose it is convenient to use an integral of that Maxwell equation concerned with curl **E**, which is slightly more general than that given by Stokes theorem. Stokes theorem itself tells us that for any closed curve, lying wholly or partly in the matter of stationary or moving bodies or in empty space

$$\int\mathbf{E}\cdot d\mathbf{s}=-\int\int\frac{1}{c}\frac{\partial\mathbf{B}}{\partial t}\cdot d\mathbf{S} \tag{13}$$

In equation 13 the integral on the left side is taken round the closed curve. The integral on the right side is taken over any 2-sided surface bounded by the closed curve, and lying where it may, wholly or partly within stationary or moving matter.

Equation 13 refers to any particular closed curve with any bounded surface which we may arbitrarily select or construct at any particular instant of time. Let us now consider the temporal succession of such closed curves and bounded surfaces which we obtain by assigning an arbitrary continuous velocity **u** to the various points of the closed curve and bounded surface. **u** need have no relation to the velocity **v** of the matter through which the curve or surface may pass. **u** is a mathematical parameter arbitrarily assigned over the curve and surface of integration in matter or in empty space, wherever the curve or surface may lie.

At any particular instant of time, equation 13 applies for that particular instantaneous configuration of curve and surface of integration. Now add to both sides of equation 13 the quantity $\int\frac{1}{c}[\mathbf{u}\times\mathbf{B}]\cdot d\mathbf{s}$. We get

$$\begin{aligned}\int\left(\mathbf{E}+\frac{1}{c}[\mathbf{u}\times\mathbf{B}]\right)\cdot d\mathbf{s}&=\int\frac{1}{c}[\mathbf{u}\times\mathbf{B}]\cdot d\mathbf{s}-\int\int\frac{1}{c}\frac{\partial\mathbf{B}}{\partial t}\cdot d\mathbf{S}\\&=-\int\frac{1}{c}\mathbf{B}\cdot[\mathbf{u}\times d\mathbf{s}]-\int\int\frac{1}{c}\frac{\partial\mathbf{B}}{\partial t}\cdot d\mathbf{S}\\&=-\frac{1}{c}\frac{d}{dt}\int\int\mathbf{B}\cdot d\mathbf{S}\end{aligned} \tag{14}$$

That is, the integral of $\mathbf{E}+\frac{1}{c}[\mathbf{u}\times\mathbf{B}]$, where **E** and **B** are the one and only Maxwellian **E** and **B**, around any closed curve, the points of which move with arbitrarily assigned velocity **u**, is equal to $-1/c$ times the rate of change of the magnetic flux linked by the arbitrarily moving curve. We shall make frequent use of equation 14, so let us keep it in mind.

Let us now consider the two slide-bar problems. We shall confine our attention to the neighborhood of the plane of symmetry perpendicular to the slide bar at its middle, and we shall assume the bar long enough so that the effect of its ends will not disturb the 2-dimensional character of the field near the plane. That is, we shall assume that near the plane of symmetry, the magnetic vectors **H** and **B** are parallel to the plane, and that the electric vectors, **E** and **D**, and also **i**, are perpendicular to the plane, and that these vectors do not change in magnitude or direction as we move away in a perpendicular direction from the plane of symmetry.

Far enough away from the slide bar we assume that we have a uniform magnetic field, $\mathbf{H}=\mathbf{B}=\mathbf{B}_0$, and a uniformly zero electric field, $\mathbf{D}=\mathbf{E}=\mathbf{E}_0=0$. Within or near the slide bar, for either case, the magnetic and electric fields are altered and **B** and **E** may not be equal to $\mathbf{B}_0$ and $\mathbf{E}_0$ respectively. Consider the temporally changing rectangular path of integration constructed as follows, Figure 1.

Adjacent to or within the central part of the slide bar, draw a short path length, *AB*, parallel to the bar, and let *AB* move with the velocity **v**, keeping always in the same geometric relation to the slide bar. Draw the two path sides *AD* and *BC* parallel to each other and in the direction **v**. The final closing path side, *CD*, is parallel to *AB*, and is to be kept at rest. *AD* and *BC* are thus lengthening, and the area *ABCD* is increasing at the rate *vl* where *l* is the length of *AB* or *CD*.

Now apply the theorem of equation 14 to the closed path *ABCD*. At *CD*, we have $\mathbf{u}=0$, $\mathbf{E}=0$. *CD* therefore gives a zero contribution to the integral of equation 14. For sides *BC* and *AD*, **E** and $[\mathbf{u}\times\mathbf{B}]$ are perpendicular, and so these sides also give zero contribution. For side *AB*, which alone contributes to the integral, we have $\mathbf{u}=\mathbf{v}$, so that

$$\int_{ABCD} \mathbf{E}+\frac{1}{c}[\mathbf{u}\times\mathbf{B}]\cdot d\mathbf{s}=\left(\mathbf{E}+\frac{1}{c}[\mathbf{v}\times\mathbf{B}]\right)\cdot\mathbf{l} \tag{15}$$

where the meaning of the vector **l** is sufficiently self-obvious.

Now, since *AB* moves with velocity **v**, keeping always in the same position relative to the slide bar, the magnetic field configuration at that end of *ABCD* remains unchanging. The rate of change of the flux linked by *ABCD* will be its rate of increase of area, *vl* multiplied by $\mathbf{B}_0$, or $[\mathbf{v}\times\mathbf{B}_0]\cdot\mathbf{l}$. Applying equations 14 and 15, paying due attention to algebraic signs, we get this equation

$$\mathbf{E}+\frac{1}{c}[\mathbf{v}\times\mathbf{B}]=\frac{1}{c}[\mathbf{v}\times\mathbf{B}_0] \tag{16}$$

Now for the iron bar, we have from the constitutive equation

$$\mathbf{i}=\sigma\left(\mathbf{E}+\frac{1}{c}[\mathbf{v}\times\mathbf{B}]\right)=\sigma\left(\frac{1}{c}[\mathbf{v}\times\mathbf{B}_0]\right) \tag{17}$$

We see then that the current set flowing in the iron bar by its motion in the otherwise uniform magnetic field, $\mathbf{B}_0$, is entirely determined by its conductivity at rest and is independent of its magnetic properties. If the bar causes the magnetic field within it and in its neighborhood to be different from $\mathbf{B}_0$, then its motion **v** will cause an electric field **E** to be set up within it and in its neighborhood given by equation 16, and the current density remains given by equation 17.

We see then that Alter Ego's commutatorless d-c machines are inoperative, but we sympathize with him for being misled by the notion that there exists a "motional electric field." There is but one electric field, **E**, and Maxwell is its prophet!

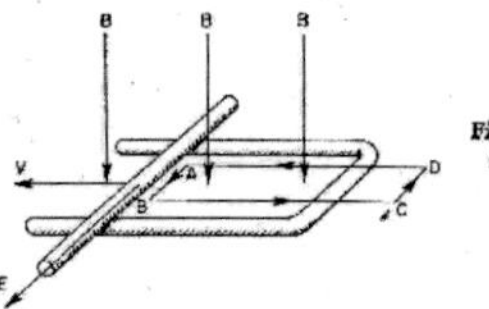

Figure 1. Path of integration, ABCD

We see also how the incorrect treatment of Lorentz' electron theory given by Jack could be corrected.

For the glass rod, if we define its polarization, **P**, by equation 18, and use the constitutive equation 1 and equation 16, we get

$$\mathbf{P}=\frac{1}{4\pi}(\mathbf{D}\quad\mathbf{E})=\frac{1}{4\pi}(\epsilon-1)\left(\mathbf{E}+\frac{1}{c}[\mathbf{v}\times\mathbf{B}]\right)=\frac{1}{4\pi}(\epsilon-1)\left(\frac{1}{c}[\mathbf{v}\times\mathbf{B}_0]\right) \tag{18}$$

The polarization then is the same as would be produced in the bar at rest by an electric field equal to $\frac{1}{c}[\mathbf{v}\times\mathbf{B}_0]$ although actually the moving glass bar will have in it a magnetic field differing slightly from $\mathbf{B}_0$ by equation 2.

J. SLEPIAN

(Westinghouse Research Laboratories, East Pittsburgh, Pa.)

REFERENCES

1. Electrostatic Space Ship, J. Slepian. *Electrical Engineering*, volume 69, March 1950, page 248.

2. Motionally Induced Electromotive Force—Part I, J. Slepian. *Electrical Engineering*, volume 69, November 1950, pages 1025-6.

3. Polarization or Charge in a Dielectric? J. Slepian. *Electrical Engineering*, volume 69, September 1950, page 792.

4. Motionally Induced Electromotive Force Part III, J. Slepian. *Electrical Engineering*, volume 70, February 1951, pages 161-2.

Three Impedances. The following is the author's answer to his previously published essay (*EE, Feb '51, p 160*).

From equations used to convert networks from star to delta, and from delta to star after substitutions and reductions, the following relationships are obtained:

$$Z_1^2=-Z_2^2=-Z_3^2 \tag{1}$$
$$Z_2=-Z_3 \tag{2}$$

These are satisfied by impedances that are equal in magnitude, one being real and positive, and the other two being conjugate imaginary impedances.

$Z_1=K$
$Z_2=jK$
$Z_3=-jK$

A. A. KRONEBERG

(Southern California Edison Company, Los Angeles, Calif.)

M.10 Motionally Induced Electromotive Force–Part V

Here Jack has difficulty with the definition of relative velocity.

Electrical Essays

Motionally Induced Electric Fields—Part V

Relative Motion of Magnetic Field and Body

Jack, the physicist, is continuing his lectures to Alter Ego and his friends on the basic principles underlying dynamo-electric machines.

Jack is trying to determine the motional electric field induced in a moving iron bar. He has declared that the motional electromotive force, or $\mathbf{E}_{mot}\cdot d\mathbf{s}$, induced in a coil which is part iron and part copper and which is being translated in an otherwise uniform magnetic field, will be zero, thus declaring inoperative some extraordinary inventions of Alter Ego. However, Alter Ego had based his expectation of a nonzero motional electromotive force in such a coil in translation in an otherwise uniform magnetic field on the equation

$$\mathbf{E}_{mot}=\frac{1}{c}[\mathbf{v}\times\mathbf{B}] \tag{1}$$

which Jack had said was universally valid. In this equation, up to this point, Jack had defined $\mathbf{v}$ as the velocity of the material of the moving body, in some suitably chosen material frame of reference.

Jack recognizes the contradiction between his statement of the universal validity of equation 1 and his statement of the inoperativeness of Alter Ego's inventions. He is now trying to resolve the dilemma by changing the meaning to be ascribed to $\mathbf{v}$ in equation 1.

Jack: "Perhaps the way out is to make $\mathbf{v}$ in equation 1 not represent the velocity of the iron bar relative to some material frame of reference which we arbitrarily say is at rest, but to make $\mathbf{v}$ represent the relative velocity of the iron bar and the magnetic lines of force.

"Thus, in Figure 1 we may regard the lines of magnetic force, A, B, C, and so forth, as at rest where they are remote from the iron bar. Then as the bar moves up to any line of force, the line of force will bulge out to meet it, as line A in Figure 1. At this bulge which is thus produced, the line of force is not at rest, but is advancing to meet the iron bar. This relative velocity of the bulge and bar increases, until the bulge hits the bar, as is shown for line B in Figure 1.

"After the line of force enters the iron, the bulge starts to straighten out, as at C. This straightening out of the bulge in the iron makes the lines of force in the iron move along in the same direction as that in which the iron bar is moving, so that the relative velocity of the bar and the magnetic lines is less than the velocity of the bar itself.

"The bulge is straightened out at D, but then develops

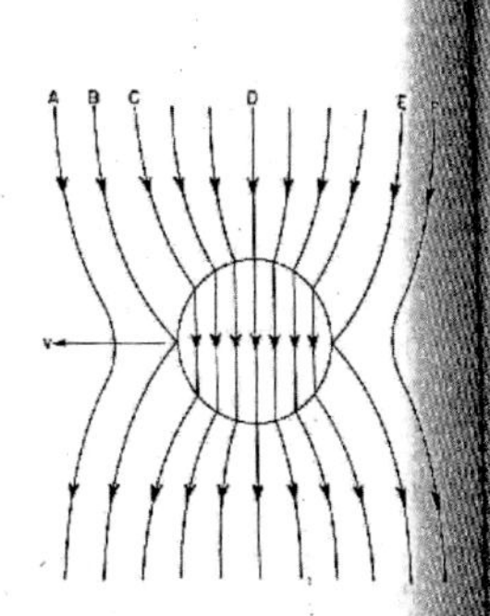

Figure 1. Converging field in a moving iron cylinder

in the opposite direction, keeping the same reduced velocity relative to the bar. At E, the line emerges from the iron, and snaps straight again through intermediate form F.

"If $\mathbf{B}_0$ is the field strength remote from the bar, where the lines are stationary, and $\mathbf{B}$ is the magnetic field strength in the iron, and if $\mathbf{v}_0$ is the velocity of the bar, and if $\mathbf{v}$ is the relative velocity of the bar and the moving field lines in it, then

$$[\mathbf{v}\times\mathbf{B}]=[\mathbf{v}_0\times\mathbf{B}_0] \tag{2}$$

since each side of equation 2 gives the rate of cutting of lines of force by the bar. Hence, the motional field in the iron will be

$$\mathbf{E}_{mot}=\frac{1}{c}[\mathbf{v}\times\mathbf{B}]=\frac{1}{c}[\mathbf{v}_0\times\mathbf{B}_0] \tag{3}$$

which is the same as that for the copper bar."

Alter Ego: "Well, if you are going to interpret $\mathbf{v}$ in equation 1 that way, I concede that it will ruin my iron-copper coil commutatorless d-c generator. But what happens to my space ship and commutatorless d-c motor? Your universal equation for the mechanical force per unit volume on a body carrying current density $\mathbf{i}$ in a magnetic field $\mathbf{B}$ is

$$\mathbf{F}_m=[\mathbf{i}\times\mathbf{B}] \tag{4}$$

"This equation does not have $\mathbf{v}$ in it at all. Therefore, the relative motion of the magnetic lines does not have any effect, so that with $\mathbf{B}$ twice as great, the force on the iron side of the coil will be twice as great as that on the copper side, and there will be a net force, not zero, tending to translate the coil."

(To be continued)

I think Jack will be able to take care of the mechanical forces. Can you, dear reader, also do so?

How about those relatively moving lines of magnetic force? Will Jack be able to get anywhere with that idea?

J. SLEPIAN
(Westinghouse Research Laboratories, East Pittsburgh, Pa.)

M.11 Motionally Induced Electromotive Force–Part VI

Dr. Slepian had a disabling stroke from which he partially recovered and lived for several more years but never published the answer to Question 2 of Part VI. The simple solution may be obtained by placing an observer on the large iron cylinder.

Electrical Essays

Motionally Induced Electric Fields—Part VI

Motional Field Inside Moving Iron Tube

Jack, the physicist, is continuing his lectures to Alter Ego and his friends on the basic principles underlying dynamo-electric machines. Jack has altered the meaning of **v** in his universal equation $\mathbf{E}_{mot} = \frac{1}{c}[\mathbf{v} \times \mathbf{B}]$ to now be the "relative velocity of the body and the magnetic field." He is now considering how to save his universal equation for the force on a current carrying body, $\mathbf{F}_m = [\mathbf{i} \times \mathbf{B}]$, when applied to an iron bar carrying current density **i** in an otherwise uniform magnetic field $\mathbf{B}_0$.

Jack: "You have forgotten, Alter Ego, the discussion we had at lunch some months ago on the forces on iron in a magnetic field (*EE, June '50, p 550*). We all agreed that a piece of iron of itself when placed in a nonuniform magnetic field would experience a force tending to move it from the weaker field towards the stronger. We had some disagreement as to just what the details of that force would be. I said that the force arose from the action of the field on the amperian currents, $\mathbf{i}_{amp} = \text{curl } \mathbf{M}$, where **M** is the magnetization density. Bill, however, said it was due to the apparent magnetic charge, $\rho_m' = -\text{div } \mathbf{M}$, being acted upon by the magnetic field **H**, and Charlie said that it was due to the magnetic doublet density **M** being acted on by the nonuniform field **H**. In any case, when we calculated the force on a total body we came to the same total result, in spite of our different equations, and we would agree that an iron body would be pulled in the direction from the weaker field to the stronger field.

"Now in the case we are considering, when we pass current through the bar, lying in the otherwise uniform field, we strengthen the field on one side of the bar and weaken it on the other. The force arising from the curr will be

$$\mathbf{F}_m = [\mathbf{i} \times \mathbf{B}]$$

and will tend to move the bar from the strengthened f to the weakened field. However, as I have just s because of the magnetization of the bar **M**, there also be a force depending on **M**, tending to pull the ba the opposite direction, namely from weak field to strc Thus, the gain in the force $\mathbf{F}_m = [\mathbf{i} \times \mathbf{B}]$ which **M** bri about because it strengthens **B** from its original va $\mathbf{B}_0$ is compensated by the tendency from the magneti bar with magnetization **M** to move from the weak fielc the strong. This force is calculable by my equation

$$\mathbf{F}_J = [(\text{curl } \mathbf{M}) \times \mathbf{B}]$$

or by Charlie's equation

$$\mathbf{F}_C = -\mathbf{H} \text{ div } \mathbf{M}$$

or by Bill's equation

$$\mathbf{F}_B = \mathbf{M} \cdot \nabla \mathbf{H}$$

"If you use any one of these equations with equatio and integrate over the whole bar, you will find that total net force remains the same as if **M** had not develop or as if the force per unit volume had remained $\mathbf{F}_m$ $[\mathbf{i} \times \mathbf{B}_0]$."

Alter Ego: "Then there was no use in my increasi **B** in my one coil side by making it of iron, since the incre in **B** only brought about a corresponding decrease in t relative velocity **v** so that $[\mathbf{v} \times \mathbf{B}]$ did not change. Ho ever, I have another form of my invention, where I redu **B** on the one coil side to practically zero by slipping hollow cylinder of extremely high permeability iron arou it, as in Figures 1 and 2. The lines of force of the magne field are completely shunted away from the coil side by t iron tube which moves along with it. Now, with yo new interpretation of **v** in your universal equation f $\mathbf{E}_{mot}$, you must make the lines of the magnetic field wh across the coil side inside of the iron cylinder with increased relative velocity **v** such that

$$\frac{1}{c}[\mathbf{v} \times \mathbf{B}] = \frac{1}{c}[\mathbf{v}_0 \times \mathbf{B}_0] \quad ($$

where $\mathbf{v}_0$ is the actual velocity of the coil side, and $\mathbf{B}_0$ th intensity of the otherwise uniform field. But with re high permeability iron, I can reduce **B** inside the iron tul to practically zero, so you will have to make **v** the relati velocity of the magnetic lines and the coil sides practical infinite if you want equation 5 to hold. I don't thin you'll like that, Doctor Jack, since you once told me th it is impossible to have a relative velocity greater than th velocity of light. If **v** has to stay finite, then with **B** equ to zero the motional electromotive force in that coil sic will be zero, and the total motional electromotive force the whole coil translating in the otherwise uniform fiel

A B C D

CYLINDRICAL IRON TUBE

COPPER COIL SIDE

Figure 1. Iron tube encircling coil side

Figure 2 (right). Coil side in iron tube in magnetic field

350 *Electrical Essays* ELECTRICAL ENGINEERIN

will not be zero, and I'll get back my commutatorless d-c generator.

"Now if I send current through the coil, the side inside the tube being in zero magnetic field will have zero force on it, whereas the other coil side will have the full $\mathbf{F}_m = [\mathbf{i} \times \mathbf{B}_0]$ on it, so now I get back my commutatorless d-c generator and space ship."

Jack: "It is true that the force on the coil side inside the iron tube will be zero, but there will be a force on the iron tube itself. When current flows in the coil, the tube will be in a nonuniform field, and it will want to move from the weaker field to the stronger. If you will use my equation, or Bill's, or Charlie's for the force on iron in a magnetic field, you will find that it is exactly equal to $[\mathbf{i} \times \mathbf{B}_0]$ integrated over the coil side. The total force on the whole coil plus iron tube will be zero, so that again your commutatorless motor and space ship are ruined. It follows then from the principle of conservation of energy that your commutatorless generator must fail too.

"But I see now that I am still not interpreting the universal equation $\mathbf{E}_{mot} = \frac{1}{c}[\mathbf{v} \times \mathbf{B}]$ rightly since I certainly do not want any infinite v. Perhaps this is the way it should be done. Perhaps we should consider how the magnetic field which acts on the moving bar originates. The original uniform field $\mathbf{B}_0$ is due to some stationary magnet or field structure and is at rest. When we use the iron slide bar, the bar becomes magnetized, and its magnetization $\mathbf{M}$ by itself would produce a field $\mathbf{B}_1$. The field $\mathbf{B}_1$, however, moves along with the magnetized bar so that its velocity relative to the bar is zero. The total field is $\mathbf{B} = \mathbf{B}_0 + \mathbf{B}_1$, but because the relative velocity of the bar to $\mathbf{B}_0$ is v_0, and to $\mathbf{B}_1$ is zero, the motional field in the bar is

$$\mathbf{E}_{mot} = \frac{1}{c}[\mathbf{v}_0 \times \mathbf{B}_0] \qquad (6)$$

"Similarly for the iron tube. The zero resultant field $\mathbf{B}$ inside the tube is really the sum of the stationary field $\mathbf{B}_0$ and an equal and opposite field, $\mathbf{B}_1 = -\mathbf{B}_0$, which is set up by the magnetization of the tube, and which moves along with the tube. Since the relative velocity of the bar to $\mathbf{B}_1$ is zero, and to $\mathbf{B}_0$ is $\mathbf{v}_0$, the motional field in the bar will still be given by equation 6.

"In the general case, we look around and examine the origins of the magnetic field $\mathbf{B}$ in which our body is experiencing its motional electric field. We may find that there is one component of field, $\mathbf{B}_1$, which arises from a current-carrying coil which is moving with velocity $\mathbf{v}_1$. The field $\mathbf{B}_1$, of course, moves along with the coil, and therefore also has the velocity $\mathbf{v}_1$. Another part of the field, $\mathbf{B}_2$, may originate in a magnetized body moving with velocity $\mathbf{v}_2$, so $\mathbf{B}_2$ also has the velocity $\mathbf{v}_2$. And so on. Now, if we have a body moving with the velocity $\mathbf{v}$ in the total field, $\mathbf{B} = \mathbf{B}_1 + \mathbf{B}_2 + \ldots\ldots$, since it is the relative velocity of the body to each component field which really counts, we have for the motional electric field in the body

$$\mathbf{E}_{mot} = \frac{1}{c}[(\mathbf{v} - \mathbf{v}_1) \times \mathbf{B}] + \frac{1}{c}[(\mathbf{v} - \mathbf{v}_2) \times \mathbf{B}_2] + \ldots$$

$$= \frac{1}{c}[\mathbf{v} \times \mathbf{B}] - \frac{1}{c}[\mathbf{v}_1 \times \mathbf{B}_1] - \frac{1}{c}[\mathbf{v}_2 \times \mathbf{B}_2] \ldots \qquad (7)$$

"This equation 7 will take care of your iron slide bar and of your iron-encircled copper slide bar, Alter Ego."

(To be continued)

Question 1. Jack's interpretation of this motional electric field equation is getting quite complicated. Will this resolving of the actual field into variously moving component fields get Jack out of his troubles?

Question 2. To calculate the force on Alter Ego's iron structures carrying current in an otherwise uniform magnetic field, Jack offered several alternative equations, saying that they each individually would lead to the same remarkable result of the cancelling out of any effect arising from the magnetization on the total force on the current-carrying structure. These individual equations would give different volume distributions of force over the structure, and it seems odd that they should all add up to the same remarkable zero resultant.

Is there not some more direct way of calculating the total force which will not inquire into these dubious volume distributions of force and which will give the total force from the externally observable fields?

J. SLEPIAN (F '27)
(Westinghouse Research Laboratories, East Pittsburgh, Pa.)

Motionally Induced Electric Fields—Part VI. The following is the author's answer to his previously published essay (*EE, Apr '51, pp 350–51*).

Question 1. No, this resolving of the magnetic field into variously moving component fields is without verifiable physical meaning and in many cases will lead to absurdities, as the next two essays will show. As was said in the preceding author's reply (*EE, Apr '51, pp 352–53*), the electric and magnetic fields in Maxwell's equations are vector fields, and nothing more; for each field, at each point of space, a direction and a magnitude, and that is all.

If in a given region we know these vector fields, we know all. We do not need to go ranging over the universe to see where our magnetic field "comes from" or originates" to be enabled to determine what happens when a local metal body moves in a locally completely observable magnetic field.

In general, a set of differential equations like Maxwell's, including the constitutive equations, has infinitely many solutions, in spite of the fact that the number of independent equations is equal to the number of independent unknowns. To determine a solution uniquely we need that are called boundary conditions. For a local region such as that in which our metal slide bar is moving, it is enough to know the electromagnetic field vectors throughout the region at one instant of time, and to know for a sufficient time earlier, and for subsequent time, the values of the electromagnetic vectors on a closed surface enclosing the region. The fact that such boundary conditions are sufficient for fixing the solution to Maxwell's equations is the whole meaning of the statement that Maxwell theory is a "local action" theory.

Now, in our case, what happens to the metal slide bar is completely determinable from the field vectors observed over an enclosing surface surrounding the conductor and inside the iron tube. These field vectors may be determined by direct observation of the forces on momentarily stationary or moving probe charges, or we may calculate them by the method described in the author's reply to Part IV of the series (*EE, Mar '51, pp 255–56*). We find, of course, that there is a much reduced magnetic field **B**, but also there is an electric field **E** observable by the force on a charged probe momentarily at rest, in the empty space inside the iron tube, and outside also, with

$$\mathbf{E}+\frac{1}{c}[\mathbf{v}\times\mathbf{B}]=\frac{1}{c}[\mathbf{v}\times\mathbf{B}_0] \qquad (1)$$

Is this field **E** a motional field, or is it induced by varying magnetic fields? That is a meaningless question.

Now, since the constitutive equation for the metal bar inside the iron tube is

$$\mathbf{i}=\sigma\left(\mathbf{E}+\frac{1}{c}[\mathbf{v}\times\mathbf{B}]\right) \qquad (2)$$

we see by equation 1 that the flow of current in it is the same as if the bar were not inside the iron tube but was moving in the uniform magnetic field $\mathbf{B}_0$ and with the electric field **E** equal to zero. We do not need Jack's device of requiring the weak field **B** to "really" be a "stationary" field $\mathbf{B}_0$, with a nearly equal and opposite field, $\mathbf{B}-\mathbf{B}_0$, "moving" with velocity **v**. There is no place in Maxwell's equations for a field velocity **v**. The place for **v** is in the constitutive equations for material moving with velocity **v**. Macroscopic material has the property of recognizable continuing individuality, so that for it the velocity **v** has meaning

Question 2. As the author's reply to question 1 is a bit long, the author asks the reader's indulgence in postponing the discussion of question 2 to the next issue of *Electrical Engineering*.

J. SLEPIAN (F '27)
(Westinghouse Research Laboratories, East Pittsburgh, Pa.)

NOTE: In the May issue of *Electrical Engineering* in the answer to Motionally Induced Electric Fields—Part VI, page 452, Doctor Slepian said that he would postpone the answer to Question 2 until the June issue. As Doctor Slepian is ill, this answer will appear in a later issue.

Essays P. 540 JUNE 1951 ELECTRICAL ENGINEERING

M.12 Calculation of induced voltages in metallic conductors by Herbert Dwight

Here is a paper published about 25 years after Albert Einstein explained the special relativistic effects[18] produced when material media travels through electromagnetic fields. Dwights rule for moving contacts and conductors of finite cross section is not correct. It appears that electrical engineers' knowledge of EMF never entered the twentieth century. The reader is warned that many modern textbooks still cite Dwight whilst attempting to explain the calculation of EMF. Included at the end of the paper are discussions of several scientists and engineers, including Dr. Joseph Slepian. Dr. Bewley, also cited in many modern texts joins in the discussion. It is clear that he also does not really understand the calculation of EMF. The erroneous papers in the electrical engineering and physics journals of the first three decades of the 20th century along with the lack of knowledge exhibited by various Westinghouse engineers and scientists probably caused Dr. Slepian to write Essays on EMF. His essays undoubtedly were the inspiration for my colleagues and me to write this book.

Calculation of Induced Voltages in Metallic Conductors

BY HERBERT BRISTOL DWIGHT*
Fellow, A. I. E. E.

Synopsis:—*A review is given of the rules and procedure used in electrical engineering for the calculation of induced voltages in metallic conductors, not including radiation effects. For the e. m. f. in a closed circuit $e = -d\varphi/dt$, the restriction is stated that the circuit is a closed loop of infinitesimal cross section and without parallel branches. More complicated circuits are to be calculated by replacing them by a number of elementary circuits. Where there are moving contacts and conductors of finite cross section, the rule $e = B\,l\,v$ should be used. A discussion is given of calculations by considering magnetic fields to have velocities and the restriction is stated that, in general, the resultant magnetic field should not be used with the $B\,l\,v$ rule, when it is made up of component fields having velocities which differ in magnitude or direction.*

* * * * *

A recapitulation of rules and methods for calculating induced voltages in metallic conductors, not including radiation effects, is desirable for use with the variety of problems encountered in electrical engineering. A careful wording of these rules, made as complete as possible, is needed in order to make them consistent.

It is believed that this paper does not present any general methods or rules for calculating induced voltages in conductors not already used by authoritative writers on electrical calculations. There are presented in this paper, however, certain restrictions which it is believed should be given along with the different rules for calculating induced voltages. These restrictions are moderate. The standard according to which they have been drawn up is that when a rule is found given in a textbook with certain wording, and when the plain, every-day meaning of the English wording can be applied to a definite problem so as to obtain an incorrect result, then a restriction is needed on the application of the rule or in the definition of its terms.

In order not to try to take away unnecessarily, from electrical engineers, useful and safe working tools, the writer has in no case stated that a rule must be discarded. It is believed that moderate restrictions are all that is required.

If it is believed by others that the restrictions presented herewith are not severe enough and should be made more sweeping, all that is required is for them to describe a practical and definite problem and show that the plain meaning of the rule with its moderate restriction gives a wrong result. This would probably still not be evidence that the rule should be discarded, but merely that the restriction should be re-worded.

It is not right for any one to say that a certain rule used by authoritative writers should be completely given up, merely because he has a more complicated alternative method of making the same calculation and so never needs to use the rule. It should be shown in addition that the rule is unreliable in definite cases, and that it is apparently impossible to make it reliable by re-wording it or restricting it.

For example, many books on elementary physics and electrical engineering give no hint as to how to calculate the e. m. f. induced in a short length of straight conductor by neighboring a-c. circuits, and yet such a calculation is described in other elementary books and the result is obtainable according to modern electromagnetic theory. As another example, many books do not mention moving magnetic fields, and yet many others describe the induced e. m. f. of an a-c. generator as caused by the magnetic lines of force of the poles moving along with the poles as they revolve.

The writer who would state that either of the above well-known rules should never be used by electrical engineers is attempting something much more radical than the mere statement of a moderate restriction. The rules are in sufficiently common use that a book which purports to give a complete statement of the calculation of induced e. m. fs. for electrical engineers should give a recommendation for or against their use. When such a book does not mention a certain rule used by engineers, one does not know whether to interpret it as a recommendation not to use the rule, or otherwise. The question of where the rules are reliable should be faced and not dodged.

There are many problems handled by the engineers by their working rules, which the physicists would find inconvenient to solve by Maxwell's equations unless they first proved some working rules, the same as or similar to the ones now in use. The physicists would do a real service to the engineers if they would prove some working rules and advocate them. This would automatically bring out the necessary restrictions.

According to Faraday's law, if there is any closed linear path in space and if the amount of magnetic flux surrounded by the path varies with time, then an electromotive force is induced around the path, which is equal in amount to the negative rate of change of the flux φ in lines or maxwells per second. Thus,

*Professor of Electrical Machinery, Massachusetts Institute of Technology, Cambridge, Mass.

Presented at the Winter Convention of the A. I. E. E., New York N. Y., Jan. 27-31, 1930.

$$e = -\frac{d\varphi}{dt} \text{ abvolts} \qquad (1)$$

where t is time measured in seconds.

The closed linear path or circuit is the boundary of a surface through which the magnetic flux passes. It is a geometrical line and has length, but no finite thickness. Since it is the boundary of a single area, it cannot have branches in parallel. At the time considered, it can be changing in shape or position. However, if it changes its position, this is done by moving from one position to the other and traversing the intermediate space.

If a loop of wire of negligible cross section occupies the same place, and has the same motion, as the path in space just considered, the electromotive force $-\frac{d\varphi}{dt}$ tends to drive a current of electricity around the wire. This electromotive force can be measured by a voltmeter or galvanometer connected in the loop of wire. As with the path in space, the loop of wire is not to have branches in parallel if the e. m. f. is to be given by Equation (1). In general, if the wire loop is not a closed linear circuit, of infinitesimal cross-section and without branches in parallel, it is not to be assumed that the e. m. f. in it will be given directly by Equation (1). The problem must be solved by starting with elementary parts to which the fundamental equations apply, and putting the results together as the conditions of the problem demand.

If a closed wire loop has branches in parallel, the problem of calculating the electromotive force indicated by an instrument is more complicated than the calculation of an e. m. f. only, and involves the resistance of the branches and the determination of circulating currents which in general will flow in them.

A metallic circuit having finite cross-section is equivalent to a number of filamentary circuits in parallel and for accurate results, the simple Equation (1) cannot be used directly. In some cases, a finite conductor can be divided into filaments, the induced e. m. f. and resistance of each of which can be computed. From these the average induced e. m. f. along the finite conductor can be determined. Such a calculation usually includes finding the current in each infinitesimal filament, and the total resistance loss. The problem will be recognized as being a skin effect or proximity effect problem. To assume that the current is uniformly distributed over the finite cross-section, and then to compute the average induced e. m. f., is only a partial solution, and one which is applicable only at practically zero frequency. Calculations of inductance using geometrical mean distances are of this zero-frequency type.

If a coil has N turns of wire in series closely wound together so that the cross-section of the coil is negligible compared with the area enclosed by the coil, or if the flux is practically all confined within an iron core and so is enclosed by all N turns alike, the e. m. f. induced in the coil is

$$e = -N\frac{d\varphi}{dt} \text{ abvolts} \qquad (2)$$

In such a case, $N\varphi$ is called the number of interlinkages of lines of magnetic flux with the circuit.

The effect of the cross-section of a coil of fine wire is taken into account if φ is taken equal to the average amount of flux surrounded by the various turns. It is to be noted that the average e. m. f. per turn in a coil made up of fine wires or filaments is the same whether the filaments are in series or parallel only if the filaments all carry equal currents. This provision is not true except at zero frequency for the case of filaments in parallel, that is, the case of conductors of large cross-section.

The minus sign in Equation (1) denotes that the direction of the induced e. m. f. is such as to produce a current opposing the change of flux. If the flux is changing at a constant rate, the e. m. f. in abvolts is numerically equal to the increase or decrease in lines in one second.

The change in magnetic flux referred to in Equation (1) may be due to a variety of causes: it may be due to relative motion between the coil and the portion of apparatus which causes the magnetic flux, as in a rotating field generator; it may be due to a change in the reluctance of the magnetic circuit, as in an inductor type alternator; it may be due to changes in the primary current producing the flux, as in a transformer; it may be due to variations in the current in the same coil in which the voltage is being induced, as in the case of self-inductance; or it may be due to change in shape occurring in a flexible secondary coil.

The restrictions given in this paper as to what is meant by a linear path or circuit for use with Equation (1) preclude the possibility of changing the amount of flux through the circuit by connecting on some parallel branches and disconnecting others. This, of course, does not induce an e. m. f.

When the flux φ surrounded by a loop of fine wire varies with the time according to the sine law, we have

$$\varphi = \varphi_m \sin 2\pi f t \qquad (3)$$

where φ_m is the maximum value of the flux and f is the frequency in cycles per second. From this we obtain by Equation (1),

$$e = -\frac{d\varphi}{dt} = -2\pi f \varphi_m \cos 2\pi f t \text{ abvolts} \qquad (4)$$

In problems involving sinusoidal alternating currents, it is thus not necessary to calculate the rate of change of flux in a loop, but merely the maximum value of flux surrounded by the loop. This simplifies the calculations to some extent.

In making calculations of flux, use is often made of

Ampère's law that a magnetomotive force is produced around any closed linear path, equal to 4π times the current surrounded by the path, absolute units being used. Here, the linear path is a geometrical line and the current may have a finite cross-section. If a condenser is included in the electric circuit, as in Fig. 1, the current i supplies charges q and $-q$ to the plates of the condenser. Then, if the conductivity of the dielectric is negligible, and if appropriate units are used,

$$i = \frac{dq}{dt} = \frac{1}{4\pi}\frac{d\psi}{dt} \tag{5}$$

where ψ is the dielectric flux passing from one plate to the other. A magnetomotive force is produced around a closed linear path, proportional to $\frac{d\psi}{dt}$, where ψ is in this case the dielectric flux surrounded by the linear path. The quantity $\frac{1}{4\pi}\frac{d\psi}{dt}$ is called the displacement current. If a surface is taken of which the linear path is the boundary, the total m. m. f. around the linear path is the sum of the m. m. fs. produced by the

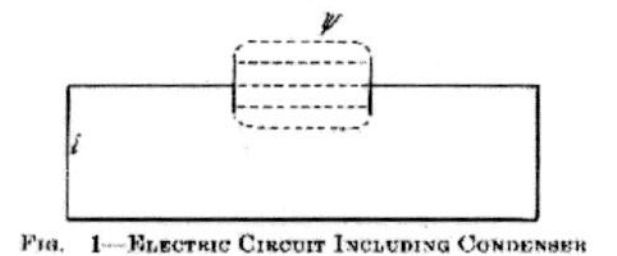

FIG. 1—ELECTRIC CIRCUIT INCLUDING CONDENSER

rate of change of dielectric flux ψ crossing the surface and by the current i crossing the surface.

While the e. m. f. e is equated to $-\frac{d\varphi}{dt}$ in Equation (1), the two phenomena do not occur at exactly the same time. The e. m. f. at a distant place occurs after a lapse of time sufficient to allow the effect to travel from the first place to the second at a finite high velocity. There is a similar velocity in the case of m. m. f. produced by current. Such a velocity is involved in the calculation of radiation effects.

It is frequently convenient in electrical engineering to calculate the e. m. f. induced in a short length of conductor. It may be mentioned that such a result is obtainable according to modern electromagnetic theory, including cases of short conductors in the neighborhood of a-c. circuits. If a closed loop of very fine wire having no parallel branches, as described for Equation (1), is moved through a non-uniform magnetic field, which is not changing with time, the only way in which the loop of wire can surround additional magnetic flux is by cutting across magnetic lines of force. The induced e. m. f. around the loop of wire is equal to the net number of lines of force which have been cut across in such a way as to bring them inside the loop of wire. If there is assigned to each part of the wire an e. m. f. equal to the number of lines per second which have been cut across by that part of the wire, the familiar rule is obtained:

$$e = B\,l\,v \text{ abvolts} \tag{6}$$

where B is the magnetic flux density in lines per sq. cm. at the location of the part of wire considered, l is the length of the part of wire in centimeters, v is the relative velocity between the portion of wire and the magnetic flux and where the directions of B, l, and v are perpendicular to one another. If B, l, and v are not mutually perpendicular to one another, their projections at right angles to each other are to be multiplied together so that e is equal to the number of magnetic lines of force which cut the portion of wire per second. The wire is considered to be of infinitesimal cross-section, and the portion is sufficiently short that it can be considered straight and the magnetic field adjacent to it, uniform.

The following discussion gives a justification for the assigning of e. m. fs. to the parts of a loop of metallic wire which are cutting corresponding amounts of magnetic flux. It may not necessarily apply to cases where components of magnetic fields are considered to have different velocities.

Let a straight portion of wire of length l lie in a uniform magnetic field B, considered stationary, whose direction is perpendicular to the wire. Let the wire carry equal positive and negative charges, $+e$ and $-e$ abcoulombs per cm. If the wire is given a motion sidewise at a velocity v cm. per sec. in a direction perpendicular to the wire and to the field, the moving charges are the same as electric currents. Considering the positive charges first, the total positive charge is $e\,l$, the corresponding current is $e\,l\,v$, and the force exerted by the magnetic field is $B\,e\,l\,v$ dynes along the wire. The force acting on each unit positive charge is $B\,v$ dynes. Similarly, there is a force acting on each unit negative charge, of $B\,v$ dynes in the opposite direction.

Where there is a force of $B\,v$ dynes in one direction on each unit positive charge and an equal force in the opposite direction on each unit negative charge, absolute electromagnetic units being used, there is by definition an electric field of $B\,v$ abvolts per cm. The induced voltage between the ends of the portion of wire of length l is $B\,l\,v$ abvolts.

Observations on flexible conductors and on streams of electrons show that the force exerted on a current or current-carrying conductor by a magnetic field, acts on the part which lies in the magnetic field. When this force appears in the form of an e. m. f. induced in a conductor, as described in the preceding paragraph, the e. m. f. is induced in the part of the conductor which lies in the magnetic field, and which moves with respect to the field. See "The Theory of Electrons," by H. A. Lorentz, 1928 Ed., p. 15, par. 8.

The application of Equation (6) can be extended to other cases than that described by considering that magnetic flux can have a velocity. In electrical engineering, it is customary to consider the magnetic lines of force produced by an electromagnet or other magnet as moving along with the magnet when it is moved to new locations. For example, the magnetic flux associated with the poles of the rotating field of an a-c. generator is usually spoken of as cutting the stationary armature conductors and inducing voltage in them. It may be mentioned that it is frequently a help to deal with a complete magnetic circuit instead of only a portion of it.

A further extension of this general method can be made by using Equation (6) to compute the voltage induced in a short length of conductor by a short element of alternating current,[1] both the conductor and the element of current being of infinitesimal cross-section. According to the Ampère rule the magnetic field in air due to a short element of current lies in circles around the straight line in which the short element lies, radiation effects not being included. The strength of the field[2] at a point P is

$$\frac{i\,d\,l \sin\theta}{r^2} \qquad (7)$$

where $d\,l$ is the length of the element of the conductor which carries the current of i abamperes, r is the distance from the point P to the element of current, and θ is the angle between the lines $d\,l$ and r. A rule can be used that when the current dies to zero, the circular lines of force collapse, each in its own plane, and in doing so each line cuts any conductors extending through that plane and induces voltage in them. For example, in connection with Equation (98), in Scientific Paper No. 169 of the Bureau of Standards, Washington, D. C., by E. B. Rosa and F. W. Grover, there is the following sentence: "The mutual inductance of two parallel wires of length l . . . . is the number of lines of force, due to unit current in one, which cut the other when the current disappears." (See also p. 77 of "Theory of Alternating Currents" by Alexander Russell, Vol. 1, 2nd Ed.)

If the current varies as a sine wave, the total number of lines φ_m which cut a conductor in a quarter cycle should be computed by integrating maximum flux densities from the conductor outward. The maximum voltage will be given by Equation (4). If effective flux is calculated from effective current, effective volts will be obtained.

A restriction is to be applied in dealing with moving magnetic lines of force when using Equation (6). The magnetic field is not to be the resultant of component magnetic fields having velocities in different directions or of different magnitudes, if the use of the resultant field does not give the same result as is given by the use of the component fields in the calculations. When the component fields are used, the e. m. fs. from them are determined separately and then combined.[3]

A simple example showing the need for this restriction when magnetic fields are considered to be in motion, is depicted in Fig. 2. A wire A lies between two north poles of equal strength. The poles are moving in opposite directions, as indicated by the arrows, and each pole is considered to carry its own magnetic lines of force with it. The poles will induce e. m. fs. in wire A in the same direction, and the total e. m. f. will be twice as great as that due to either pole alone. The resultant magnetic field at the location of A is practically zero, and the correct e. m. f. induced in a short portion of A

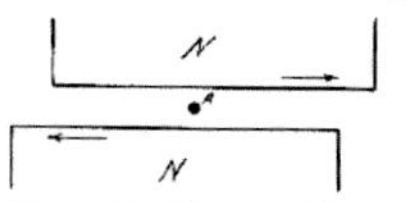

FIG. 2—WIRE IN ZERO RESULTANT MAGNETIC FIELD

cannot be calculated by using the value of the resultant field. This is evidently a case where the e. m. fs. due to the component fields must be computed separately and then combined.

It may be mentioned that a small block of iron placed in the air-gap close to A will not change the voltage which is being induced in A, if the block of iron is in a location where the resultant magnetic field is zero. The permeability of the iron depends on the resultant magnetic field.

Another example is the case illustrated in Fig. 3, where alternating currents I and $-I$ are carried by two thin tubular conductors centered on the same axis. A wire A is placed at the axis of tubes and is insulated from them.

Dividing the current I into small elements of current each equal to i, the e. m. f. induced in A by each elementary current is the same and is

$$j\,\omega\,2\,i \text{ logn } (u/a)$$

counting flux up to a certain large distance u, and writing logn to denote the natural logarithm. The quantity $\omega = 2\,\pi\,f$, where f is the frequency. All such e. m. fs. are to be added together, giving a total

$$j\,\omega\,2\,I \text{ logn } (u/a)$$

Similarly, the e. m. f. induced in A by $-I$ in the larger tube is

$$-j\,\omega\,2\,I \text{ logn } (u/b)$$

Adding the two e. m. fs. together, gives

$$j\,\omega\,2\,I \text{ logn } (b/a) \text{ abvolts per cm. of } A \qquad (8)$$

1. "Elements of Electromagnetic Theory," by S. J. Barnett, p. 337.

2. "Elements of the Mathematical Theory of Electricity and Magnetism," by J. J. Thomson, 4th Ed., Art. 211.

3. "Induced Voltages in Conductors," by H. B. Dwight, *Electric Jl.*, April 1928, p. 174.

No matter how large a value of u might be taken, the result would be the same.

The same result would be given by Equation (1) if wire A as well as the two tubes, which are shown in cross-section in Fig. 3, were extended around so as to make closed circuits.

It is to be noted that wire A lies in a region of zero resultant magnetic field and therefore the value of the voltage per cm. in A given by (8) cannot be calculated from the value of the resultant magnetic field in which wire A lies.[3]

Since an e. m. f. is induced in each centimeter of wire A, as shown by (8), the region inside the tube, although it has zero resultant magnetic field, is not in as quiescent a state as a region remote from all currents, where no voltage would be induced in a wire.

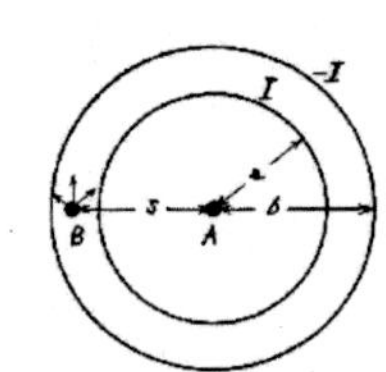

FIG. 3—WIRE IN CURRENT-CARRYING TUBES

If a parallel wire is outside the smaller tube as in the position B, Fig. 3, the drop per cm. due to I is

$$j\,\omega\,2\,I\,\log n\,\frac{u}{s}$$

and the drop due to $-I$ is

$$-j\,\omega\,2\,I\,\log n\,\frac{u}{b}$$

using the expressions for geometrical mean distance from a point to a circle excluding and including the point. The total drop is

$$j\,\omega\,2\,I\,\log n\,\frac{b}{s}$$

This value is also obtained by integrating the total lines of the resultant field from B to the outer circle. The component fields due to the filamentary currents making up I in the inner tube are not all in the same direction, as is indicated by the three arrows pointing upward from B. The component fields due to the elements of current making up $-I$ in the outer tube point in every direction from B. If the wire is outside of both tubes, the e. m. f. is zero and the resultant field is zero. Evidently, when a wire is outside the inner tube, the e. m. f. is correctly calculated from the resultant magnetic field, but when the wire is inside the inner tube, the resultant field at the location of the wire cannot be used in calculating the e. m. f. in a short portion of the wire. The criterion is the calculation based on the fundamental expression for filamentary currents.

In the class of cases where there are sliding and moving contacts between conductors of finite cross-section, as in d-c. generators and motors, and in homopolar generators, of which the well-known Faraday disk is an example, Equation (6) should be used to compute the induced e. m. f. per cm. for the various parts of the

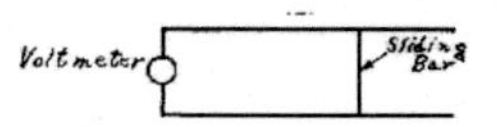

FIG. 4—SLIDING CONTACTS WITH INFINITESIMAL CONDUCTORS

conductors. These e. m. fs. can then be summed up or integrated. In cases such as a d-c. machine or a homopolar generator, there is usually at all times a conducting path for current to flow, and this may quite properly be called a circuit, but it is not a closed linear circuit without parallel branches and of infinitesimal cross-section, and therefore it is not strictly allowable to apply Equation (1) directly to such a circuit. If such a practical circuit or current path be made to enclose more magnetic flux by a process of connecting in one parallel branch conductor in place of another, then such a change in enclosed flux does not correspond to an e. m. f. according to Equation (1), because the circuit is not the kind to which Equation (1) applies. Equations involving $d\,\varphi/d\,t$ are not to be used directly on circuits where there are sliding or moving contacts between conductors of finite cross-section or between conductors connected in parallel. Since practical conductors have a finite cross-section, and since sliding contacts in rotating machines, especially commutator machines, involve conductors connected at least temporarily in parallel, it seems advisable to adopt a practical rule not

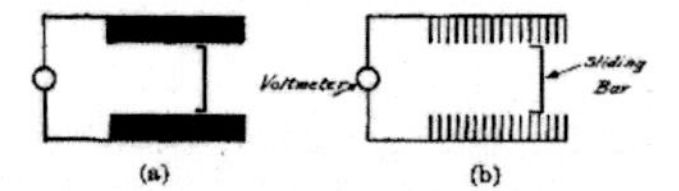

FIG. 5—SLIDING CONTACTS WITH FINITE CONDUCTORS AND PARALLEL BRANCHES

to use equations involving $d\,\varphi/d\,t$ where there are sliding or moving contacts.

A circuit containing sliding contacts which is frequently described, is shown in Fig. 4. The conductors have infinitesimal cross-section, and the e. m. f. indicated by the voltmeter can be calculated by both (1) and (6), the same result being obtained.

The case shown in Fig. 4 is often shown as a general case. In reality, it is a special case, and only such special cases involving sliding contacts can be solved by a single equation in $d\,\varphi/d\,t$. More general and practical cases are illustrated in Fig. 5, (a) and (b), where conductors of finite section and parallel branches are involved. In Fig. 5 (b) all the conductors may be considered to have infinitesimal section, but because of the parallel branches and the sliding contacts it is not possible to equate the e. m. f. to the rate of change of the flux enclosed by the path for the current. The e. m. f. is proportional to the flux swept across by the moving bar; to calculate this e. m. f. an equation such as Equation (6), which distinguishes between the moving and stationary parts, should be used.

The statements of the rules for calculating induced voltages in metallic conductors and the restrictions to be applied to them, are subject to modification as new and unusual problems are investigated; in the form given in this paper they are not arranged to include voltages produced by radiated fields. Where the word "rule" is used, a practical method of calculation is signified. This is to be distinguished from a "law." The criterion to be used regarding induced voltages is a calculation by Maxwell's equations and the extensions of them given by modern electromagnetic theory.

Discussion

Joseph Slepian: Professor Dwight is very successful in pointing out the unsatisfactory state of the theory of the induction of e. m. fs. by varying magnetic fluxes as taught to electrical engineers, but in my opinion he has been less successful in his attempt to make that theory satisfactory. The fault with the teaching of this subject to engineers, as I see it, is that engineers are not taught that varying magnetic fluxes have associated with them, or produce, electric fields in space, even at points remote from the generating magnetic fluxes. Professor Dwight makes no mention of this in his paper.

Engineers are taught that electric charges produce electric fields at points remote from the charges and where there may be no charge; they are taught that electric currents produce magnetic fields at remote points where there may be no electric current. But they are not taught at all that varying magnetic fluxes in the same way and in the same sense produce electric fields in space even at points remote from the generating varying magnetic flux and where there may be no magnetic field.

Professor Dwight in his paper does not define the e. m. f. which he is attempting to calculate. Where the definition of e. m. f. is given with any precision, it is usually this: the e. m. f. between two points taken along some path joining the points is the integral of the electric force taken along that path. The electric force itself, at any point, or the electric field intensity at that point, is defined as the force which acts upon a unit positive charge placed at the point in question, and *at rest there.* If the unit charge is moving, then in addition to the force due to the electric field, there is another force upon the charge, whose magnitude is given by the product of the velocity of the charge, the intensity of the magnetic field at the point, and the sine of the angle between the directions of the velocity and the magnetic field intensity.

This statement shows that there is a kind of relativity about the electric field, because the electric charge which is at rest relative to one observer, will be moving relative to another. The first observer ascribes all the force acting upon the charge as due to an electrostatic field. The second observer attributes some or all of the force acting upon the charge, which is moving relative to him, as due to the reaction of the magnetic field at the point in question upon the moving charge. To determine the electric field at the point, the second observer will observe the force upon a unit charge at rest relative to himself, but for the first observer, this charge will be moving, and he will in his turn ascribe some or all of this force to the reaction of the magnetic field upon the motion. Thus the two observers "see" different electric fields, for "seeing" electric fields means merely observing the motions, or the forces upon charged material bodies, and making deductions therefrom with an arbitrary assumption as to what body is at rest. Thus there will be different electric fields for different observers, and these electric fields will differ by the vector product of the magnetic force at each point and the relative velocity of the observers. However, it is not my intention to stress this relativity of the electric field but rather to point out what is necessary to give a logical and consistent picture of the electric field to a single observer. I mention the relativity because in one of Professor Dwight's examples which I shall discuss presently, there is confusion due to lack of recognition of this character of the electric field of being relative to the observer.

Let us now consider the e. m. fs. which are induced in stationary (relative to the observer) conductors. Professor Dwight discusses two examples, illustrated in his Figs. 2 and 3. In either of these cases, a stationary electric charge placed at the point a, that is at a point of the conductor subject to the induction, will experience a force acting upon it. Therefore, there is an electric field at that point, an electric field as real and existent as the electric field emanating from an electric charge. In fact it is the only kind of electric field there is, namely one deduced from the reaction upon stationary charges. Current will flow in the conductor subject to the induction due to the action of this electric field upon the mobile charges within the conductor. There need not be any magnetic field at the conductor; the electric field alone accounts for the current which may flow. The electric field may be calculated in known ways from the varying magnetic fluxes at other parts of space, and may be said to be due to these varying magnetic fluxes. To fail to consider its existence, and to attribute the force acting on charges in the conductor as due to the superposition of the effects of infinitely many component magnetic fields having all directions, and moving in all directions, is to present not only a confusing picture but one which is logically inconsistent with the definition of the electric field intensity at a point. When this electric field is taken into account, there is no room left for the reaction of these hypothetical moving magnetic fields upon stationary electric charges.

Let us now consider the case of a body moving (relative to the observer) in an unvarying magnetic field. Initially at least, there is no electric field, for a stationary electric charge experiences no force. There is therefore no e. m. f. anywhere, not even in the moving body. The charges making up the moving body will experience forces, but this is due to their motion in the magnetic field, and not to the presence of an electric field. If the body is a conductor, the charges will move due to these forces, and if the conductor is open, charges will collect at the ends of the conductor producing an electric field according to the principles of electrostatics. Evidently the electrostatic field will develop until it just balances the reaction of the magnetic field upon the charges at each point of the moving conductor. With the development of this electrostatic field there will be e. m. fs. between various points, in the conductor as well as elsewhere, and because this electrostatic field just balances the magnetic reaction, the e. m. fs. in the conductor can be calculated from the magnetic field and the conductor motion by the usual "cutting" rule.

Professor Dwight's calculation of the e. m. f. induced in the moving conductor which he presents on page 449 illustrates the confusion of thought which results from the failure to recognize that the electric field is relative to the observer. Quoting from his paper "Where there is a force of $B\,v$ dynes in one direction on each unit positive charge and an equal force in the opposite direction on each unit negative charge, absolute electromagnetic units being used, there is by definition an electric field of $B\,v$ abvolts per cm." But, the definition of electric field referred to by Professor Dwight requires that the unit positive or negative charge which experiences the force $B\,v$ dynes shall be at rest. The charges will be at rest only for an observer moving along with the conductor. For him, and for him only, will there be an electric field, and for him there will be no reaction of the magnetic field upon the charges at rest (relative to him). For him there will be varying magnetic fields at points fixed (relative to him) in space, at the boundaries of the uniform magnetic field, and from these varying magnetic fields he can calculate the electric field which he observes.

But for the observer for whom the conductor is moving, and for whom the magnetic field is unvarying there is no electric field, but only the reaction of the magnetic field upon the charges in the moving conductor. The electric field and the magnetic reaction exist only the one for the one observer, and the other for the other observer, and not both simultaneously for either observer.

When speaking of moving lines of magnetic force it must be remembered that the lines of force do not have all the reality of material objects, but that there is much arbitrariness in how we place them. We draw them in merely as a convenience in depicting the magnetic field, and the only restrictions upon them is that their directions shall be that of the magnetic field at each point, and that they shall be so spaced that their density shall be proportional to the intensity of the magnetic field at each point. This still leaves much arbitrariness in the positioning of the lines and in ascribing continuous individuality to them. Thus for a uniform field we may imagine the lines of force executing the most complicated dance among themselves that we please, so long as we keep the lines straight and parallel and spaced with uniform density, for so long will they continue to represent correctly at each point the direction and intensity of the field. But we must not expect that this assumed motion of the lines of force gives rise or corresponds to any detectable reality, so that to ask whether these motions exist or not is to ask a meaningless question. Particularly we must not expect any reaction between these "moving" lines and stationary charges.

L. V. Bewley: My discussion of Professor Dwight's paper will be a comparison with the point of view given in my own paper, *Flux Linkages and Electromagnetic Induction in Closed Circuits.*[1] The scheme of classification is vividly represented by the "Family Tree of Electromagnetic Induction" shown in Fig. 1,

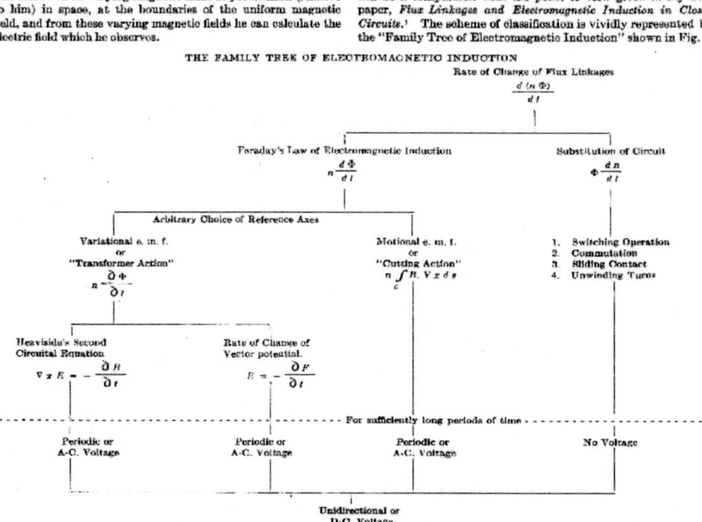

Fig. 1

the derivation of which is given in my paper. The principal features of this classification are:

1. A change of flux linkages due to a *substitution of circuit* is interpreted as $\varphi\,d\,n/d\,t$, and does not induce a voltage.

2. Faraday's Law of Electromagnetic Induction $n\,d\,\varphi/d\,t$ is universal and general. It includes any and every case of induction.

3. The occurrence of *transformer action* or of *cutting action* is entirely dependent upon the arbitrary choice of a system of reference axes. If the circuit moves as a rigid body, then by taking reference axes fixed to the circuit, the induced voltage may be computed as entirely due to the *variational component*. If the magnetic field is a rigid distribution, then by taking the reference axes fixed to this distribution, the induced voltage may be computed as entirely due to the *motional component*. If neither the circuit nor the magnetic field distribution remain rigid, then both the *motional* and *variational* components will necessarily appear in the expression for the induced voltage.

1. A. I. E. E. Quarterly Trans., Vol. 48, April 1929, p. 327.

4. The *second circuital equation* of Heaviside, and the *rate of change of vector potential*, both of which assign a definite induced voltage to each element of circuit, are specific expressions for *transformer action*. They are invoked in the computation of electromagnetic waves or radiation.

5. For sufficiently long periods of time, the induced voltage is either periodic or zero. A continuous unidirectional or direct voltage cannot be obtained (from electromagnetic induction) except through the agency of a substitution of circuit, although a substitution of circuit does not in itself generate a voltage.

General criteria for the induced voltage in any given case, and essential definitions of a *circuit* and of a *turn* will be found in my paper, as well as the derivation of the voltage induced in a circuit of any shape moving and changing shape at any variable rate through an arbitrarily varying magnetic field.

Having reviewed the general scheme of the phenomena of electromagnetic induction, let us establish the validity for computing the voltage induced in an element of wire by component traveling fields, such as illustrated in Fig. 2 of Professor Dwight's paper, or by Fig. 2 of this discussion. The flux density, expressed as a vector, is:

$$B = i\,\alpha + j\,\beta + k\,\gamma = i\,(o) + j\,[f_1\,(x + v_1\,t) + f_2\,(x - v_2\,t)] + k\,(o) \quad . \; . \quad (1)$$

where for the sake of simplicity only two waves have been taken, —one moving in the direction of positive x with a velocity v_2 and the other in the direction of negative x with a velocity v_1. The direction of B has been taken entirely along the y axis, and the conductor along the z axis.

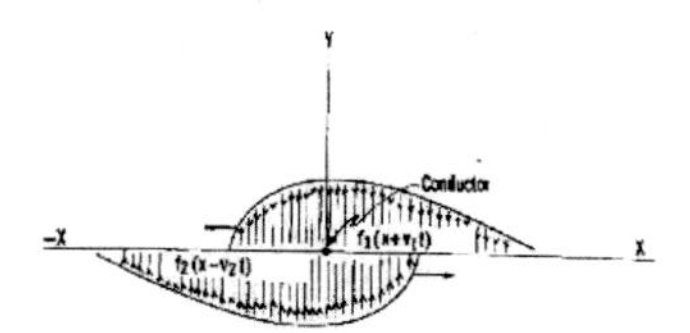

FIG 2—DISTRIBUTION OF FLUX DENSITY ALONG AND NORMAL TO THE X AXIS

Heaviside's second circuital equation is

$$\text{Curl}\,E = -\frac{\partial B}{\partial t} \qquad (2)$$

where E is the electric force and B the flux density. Expanding (2) and substituting (1) there are the three scalar equations:

$$\left.\begin{aligned} &\frac{\partial E_z}{\partial y} - \frac{\partial E_y}{\partial z} = 0 \\ &\frac{\partial E_x}{\partial z} - \frac{\partial E_z}{\partial x} = -V_1 f_1(x + v_1\,t) + V_2 f_2\,(x - v_2\,t) \\ &\frac{\partial E_y}{\partial x} - \frac{\partial E_x}{\partial y} = 0 \end{aligned}\right\} \qquad (3)$$

And these have as a solution, compatible with the implied boundary conditions,

$$\left.\begin{aligned} &E_x = 0 \\ &E_y = 0 \\ &E_z = +v_1 f_1(x + v_1\,t) - v_2 f_2\,(x - v_2\,t) \end{aligned}\right\} \qquad (4)$$

But this is the same result as is obtained by applying the (BLV) rule separately to each component flux density wave, and superimposing the results. In particular, if $f_2 = -f_1$, and if $v_2 = v_1$,

$$f_1(o + v_1\,t) + f_2\,(o - v_2\,t) = 0 \qquad (5)$$

at the conductor; although by (4)

$$E_z = v_1 f_1(O + v_1\,t) - v_2 f_2\,(O - v_2\,t) = 2\,v_1 f_1(O + v_1\,t) \qquad (6)$$

thereby verifying Professor Dwight's conclusions, and rigorously establishing the rule for voltages induced by traveling distributions of flux.

It may be remarked that a distribution of flux can be considered as moving when it is expressible as a traveling wave.

H. B. Dwight: The chief purpose of my paper is to find a wording, complete with such restrictions as may be necessary, for the well-known fundamental rules used in calculating induced voltages in metallic conductors, not including effects connected with radiation, so that a more complete statement of these rules might be included in elementary books than is usually found at present.

Faraday's law, expressed by the equation $e = -\dfrac{d\,\varphi}{d\,t}$, was deduced from observations on fine wires. Since the attempt is sometimes made to apply the law in one single calculation to circuits of heavy conductors or with branches in parallel, it seems advisable to include in the statement of the law a restriction against this.

It does not seem necessary to add a statement that electric fields are produced at points in space by magnetic fluxes which are remote from those points. This is really contained in the statement of Faraday's law, which says that an electromotive force is induced in the boundary of an area, which is, of course, separate or remote from the magnetic flux at the central parts of the area. In the paragraph of my paper following Fig. 1, the e. m. f. at a place distant from the flux is specifically mentioned.

The e. m. f. calculated in my paper is defined in the title, it is the e. m. f. electromagnetically induced in a conductor. This should be distinguished from the electric field measured by an observer with respect to whom the conductor may be moving.

Dr. Slepian's discussion regarding two observers does not apply to the rules dealt with in my paper for two reasons. First, he deals with the complete electric field at a point due to all causes. It is, however, possible to calculate the induced e. m. f. in a conductor separately and the title of the paper calls for such a calculation. Second, where only one cause is acting to induce e. m. f. in a conductor, he divides the effect into two parts, namely the electric field as measured by a stationary observer, and "another force" due to the motion of the conductor. Instead of the electromotive force in the conductor being equal to the first part, it is equal to the sum of the two parts. It should not be assumed that this indirect method is the only way to compute the induced e. m. f. in a conductor. The rules dealt with in the paper are used to a considerable extent and they take account of the motion of the conductor directly.

I wish to take exception to the statement of Dr. Slepian that there is no e. m. f. in a body moving in an unvarying magnetic field. When the directions are at right angles, the e. m. f. due to the motion is $H\,l\,v$, whether calculated by the rule of that name or by other methods. This is probably a matter of the meaning of the terms, but when positive charges are forced in one direction along a wire and negative charges in the other direction, there is an electromotive force in the wire. Further, as Dr. Slepian states, to an observer moving with the wire there is an electric field. The sentence quoted from my paper is in agreement with this. In refutation of Dr. Slepian's statement quoted in the first sentence of this paragraph, reference may be made to the second page of the paper by S. J. Barnett (A. I. E. E. TRANS., 1919, p. 1496), where in connection with a moving

conductor it is said that "the electric intensity and electromotive force in the body due to the motion have been called the motional intensity and e. m. f. by Heaviside."

Mr. Bewley's discussion tends to confirm my conclusion that an extremely short statement of rules for induced voltage, without restrictions, cannot safely be used for practical problems about electrical apparatus. This is shown by the number of items given in his table. In contrast to Dr. Slepian's statement that there is no e. m. f. in a moving body in an unvarying magnetic field, Mr. Bewley describes the motional e. m. f. in such a case.

Mr. Bewley's use of Maxwell's differential equations with the application of boundary conditions to the problem of two traveling waves, is very interesting and it is noteworthy that he does not say that all problems, short or long, should be solved by this method. The method is powerful and reliable, and I believe that electrical engineers in this country would do well to follow the European example and make more use of this method. However, it cannot be expected that the simpler rules for calculating induced voltages in conductors will be completely given up, especially as they are being taught to twelve-year-old students in Junior High Schools, including the description of moving magnetic fields, and so it seems desirable to make the wording of these rules as safe and reliable as possible.

Figure 143: Dr. Joseph Slepian

N Biography of Dr. Joseph Slepian, February 11, 1891–December 19, 1969

By T. Kenneth Fowler

JOSEPH SLEPIAN, **INVENTOR** of the ignitron and other mainstays of the electric power industry, after a lifetime career at the Research Laboratories of the Westinghouse Electric Corporation, died on December 19, 1969. Elected to the National Academy of Sciences in 1941, Slepian was a member of Section 31, now called Engineering Sciences. Indeed, his career exemplified the union of these disciplines.

Holder of 204 patents at Westinghouse, Slepian began his career as a pure mathematician. He was born in Boston on February 11, 1891, son of Russian immigrants. Advanced student status in high school allowed him to enroll at Harvard University at age 16; he made Phi Beta Kappa and received his bachelor's degree in 1911, his master's degree in 1912, and a Ph.D. in mathematics in 1913. All during this time he maintained odd jobs to help support himself, including a stint as a licensed motorman on the Boston Electric Railway.

After Harvard Slepian was able to continue a year of postdoctoral studies as a Sheldon fellow, first at the University of Gottingen in Germany and then at the Sorbonne in Paris. He returned to the United States in 1915 and accepted a position as instructor of mathematics at Cornell University. After only a year at Cornell he resigned his position to join the Westinghouse company at its East Pittsburgh Works as a student apprentice in the railway motor department. By 1917 the company had moved Slepian to the research department, shortly before the establishment of its pioneering independent research facility at Forest Hills in 1918. He advanced quickly, as head of the General Research Section in 1922, research consulting engineer in 1926, and associate director for research from 1938 until his retirement in 1956.

Slepian's move to Westinghouse proved a happy transition for all concerned, his prolific output of patentable inventions for the company being exceeded only by those of George Westinghouse himself. According to colleagues Slepian made maximum use of his mathematical talents in his new career, his inventions invariably being the consequence of careful science and theoretical analysis. In fact his talent for invention had already emerged at Cornell, where in 1915 he filed a patent for a device to measure the speed of a boat by means of magnetohydrodynamics. By 1919 he had produced his first patent at Westinghouse, for circuit interrupters. He was still pursuing inventions when I first met him, late in his career, when he was developing a new plasma method of isotopic separation, his ionic centrifuge that he pursued before and after retirement, with 20 publications on this topic alone, most of them in the Proceedings of the National Academy of Sciences.

Slepian's first major success at Westinghouse led to the autovalve lightning arrester, at a time when the cost and maintenance of conventional electrolytic arresters no longer served the needs of a growing industry. His pioneering research on lightning arresters began in 1920, three years after his transition to a career of engineering research. Characteristically, when presented with the problem, Slepian first conducted a thorough analysis of the operation of electrolytic lightning arresters during a discharge—research that disclosed the need for surge protection that he would solve with a countervoltage produced by a glow discharge in air. This in turn led to his many experimental and theoretical contributions to the field of electrical conduction through gases and a familiarity with plasma physics that inspired even his later work on the ionic centrifuge . Slepian's careful study of ionized gases also prepared the way for other notable inventions, including the deion circuit breaker and the workhorse ignitron mercury rectifier familiar to me from my earliest contact with laboratory experiments on plasmas.

Slepian's contributions to the ignitron followed a pattern established in his work on arresters. Though already commercial, mercury rectifiers had reached an impasse: unacceptable "arc-backs" that required deep analysis to unravel. Slepian provided this analysis, leading him to propose separating the multiple rectifier anodes into individual chambers, which was the first step toward the ignitron design. There followed an intensive period of research to provide a means of extinguishing and then initiating anew the mercury arc on each operation cycle, dependably, when required, without appreciable time lag. More than 4 million kVA in ignitrons had been installed by the late 1940s.

The deion circuit breaker was also the result of detailed scientific research, in this instance on the nature and origin of arcs. As in his other research his work always involved observation and experiment as well as theory. Though first a mathematician Slepian had also become a productive and careful experimentalist in the laboratory. It was he who discovered plasma arc regimes not requiring thermionic emission of electrons from the cathode, at gas densities well below those thought possible before his work on cold-cathode heavy-current arcs. The practical result was the deion circuit breaker, employing the cold-cathode technique together with ingenious annular electrodes and voltage distribution that avoided thermionic hotspot emission that would otherwise spoil the almost instantaneous buildup characteristic of the cold-cathode operating regime.

A highlight in Slepian's career was his receipt of the Edison Medal in 1947, in part for his inventions of the autovalve lightning arrester, deion circuit breaker, and ignitron cited above. Marveling that a one-time pure mathematician would receive an award honoring a man like Edison, Slepian in his acceptance speech reflected wisely on the productive interplay of mathematics, science, and engineering. It is appropriate to quote here excerpts from his remarks, published in full in Electrical Engineering (67[1949]:258-61).

That a man with my particular kind of talents, abilities and personality should win a high engineering honor may seem very remarkable..... The dominant interest of my youth, and the kind of formal education it led me to acquire, certainly did not presage distinction in such fields.

I have pondered on what rightly may be called the really distinctive features of the mathematician, scientist and engineer. There seem to be two ways of logically distinguishing among them. One . . . is by the kinds of skills they display . . . their crafts. The other, and which I think strikes deeper, is by their motivations or compelling interests.

Let me proceed then to ask these questions. When the mathematician is doing that which is uniquely mathematics, and cannot possibly be said to be physics, or chemistry or other science; when the physicist, as a typical scientist, is doing that which is uniquely physics and cannot be said to be mathematics or engineering; when the engineer is doing that which is certainly engineering; what are their respective distinctive motivations and compelling interests?

My answers lead to the following definitions.

The "mathematician" is one whose interests and activities lie in determining and studying how things may fit together, that is, what are possible systems of order, and what are the details of such possible systems of order.

The "scientist" is one whose interests and activities lie in determining what is the actual order of things in the physical world and studying the details of that order.

The "engineer" . . . is one whose interests and activities lie in devising, designing, constructing or controlling the operation of physical devices, machines, technical processes, or services which have practical utility . . . [making] use of the accumulated knowledge, skills and techniques of the "mathematician" and the "scientist."

We know now that while "mathematicians" and "scientists" carry on their activities for "their own sake," that is for aesthetic reasons or other intellectual satisfactions, nevertheless, their work will lead to radical and revolutionary advances in technology in the future. The invention of the number system in which all algebraic equations, including $x2 + 1 = 0$, have solutions, had to be done by the "mathematician." The "engineer" could not anticipate its utility for solving practical A-C problems. Only a "physicist" would be engrossed with the faint glows given off by certain rare minerals. How would the "engineer" know that these faint glows were the indications of tremendous technically utilizable forces within the atom?

With these examples before us, we see that while there are also other important reasons, we must support "mathematics" and "science" in the United States because of the inevitable future advances in technology which they will induce. To make "mathematics" and "science" flourish, we must create for "mathematicians" and "scientists" a favorable atmosphere.

High above all other requirements in the favorable atmosphere is that of freedom; freedom to choose their work or object of interest, freedom to write and publish, freedom to communicate with their fellows.

Slepian found his own favorable atmosphere at the Westinghouse Research Laboratories, a model for other corporations at the time the Forest Hills laboratory was created. "Sometimes," he wrote for his twenty-fifth class reunion at Harvard, "I look with envy at the apparently more leisurely and less harassed lives of acquaintances in university circles, and at one time I nearly changed over to this field, but on the whole I think I am in the work and place best suited to me . . ."

Besides his own research and inventions Slepian was a valued consultant, much sought for his advice by others in the company. During World War II he both participated in the Manhattan Project and served as consultant to the Office of Scientific Research and Development, and as a dollar-a-year man with the War Production Board.

Despite leaving academia as a profession Slepian never lost interest in teaching, fulfilled at Westinghouse by his own initiative in organizing informal courses on a variety of topics, including vector analysis, the theory of electricity and magnetism, the kinetic theory of gases, and the conduction of electricity through gases. In addition, besides practical inventions, in 1922 he filed for a patent, issued in 1927, for the idea of accelerating electrons by magnetic induction, later employed in the betatron accelerator developed by Donald Kerst at the University of Illinois and used widely in nuclear physics research.

Slepian published 121 technical papers, articles, and essays, some of which are listed below. In 1933 Westinghouse Electric Company published his book, Conduction of Electricity in Gases, a compilation of his lectures for Westinghouse colleagues. This book became a classic, used by physicists and educators throughout the world.

In addition to receiving the Edison Medal in 1947, Slepian was the recipient of the John Scott Medal at the Franklin Institute in 1932 and the American Institute of Electrical Engineers' Benjamin Garver Lamme Medal in 1942. He was elected a fellow of AIEE in 1927 and the Institute of Radio Engineers in 1945 (predecessors of the Institute of Electrical and Electronic Engineers). He received the Westinghouse Order of Merit in 1935. In 1939 his scientific contributions were recognized by the French with the title Officer de Academie. In 1949 he was awarded an honorary doctor of engineering degree by Case Institute of Technology (now Case Western Reserve) and in 1955 an honorary doctor of science by the University of Leeds.

At the Lamme medal ceremony L. W. Chubb, then Westinghouse director of research, painted a charming picture of the young mathematician turned engineer.

When he first arrived at the engineering laboratories, I happened to be in charge and in a position to recognize his unusual qualities.

On one occasion the rest of us were so busy on some development that I could not assign Doctor Slepian to a new job at the moment. Instead of marking time until we finished, Slepian asked my permission to study a complicated setup of large motors, electrolytic condensers, reactors, instruments, transformers and disorderly wiring and cables in a nearby room. Permission granted, he traced the circuits and made a complete schematic diagram of the system on a large piece of paper. He did not recognize the electrolytic condensers, and I explained them to him.

Without further assistance, he deduced that the setup was designed to explore operation of polyphase induction motors from a single-phase power line. He not only learned about this specific problem but went on from there, in a short period analyzing the general problem of phase conversion, and making several inventions for both static and rotating phase splitters. His initiative and independence have not lessened during the years since then.

Slepian suffered a stroke in 1951, but though handicapped by health problems, he continued at the Westinghouse laboratory until his retirement on February 28, 1956. In private life he loved music, art, and literature. He was a season ticket holder for the Pittsburgh Symphony Orchestra for over 40 years. He liked to joke, and his friends enjoyed his incisive sense of humor.

He married Rose Myerson in 1918. They had two sons, Robert and David, both of whom followed with distinction in their father's footsteps, Robert at Westinghouse and David at Bell Laboratories. David was elected to the National Academy of Sciences in 1977.

N.1 Acknowledgements

I WISH TO THANK Joseph Slepian's son David for his help and comments. I also gratefully acknowledge the help of John Coltman, who was well acquainted with Joseph Slepian at Westinghouse, and F. A. Furfari for his help in resolving several questions. I have drawn liberally from Furfari's biographical article about Slepian in IEEE Industry Applications Magazine (November/December 2000) and from published comments by M. W. Smith at Slepian's Edison Medal ceremony and L. W. Chubb at his Lamme medal ceremony.

N.2 Bibliography of some of Joseph Slepian's papers

1919 The flow of power in electric machines. Electr. J. 16:303-11.

1920 Reactive power and magnetic energy. Trans. AIEE 39:1115-32.

1921 Why high frequency for radiation? Electr. J. 18:129-31.

1923 Surges on power systems. Electr. J. 20:176-81.

1926 Theory of current transference at the cathode of an arc. Phys. Rev. 27:407-12.

Thermionic work function and space charge. Phys. Rev. 27:112(A).

1929 With R. Tanberg and C. E. Krause. New valve-type lightning arrester, Electr. World 94:1166-67.

Theory of the deion circuit breaker. Trans. AIEE 48:523-27.

1930 Theory of a new valve type lightning arrester. Trans. AIEE 49:257-62.

1931 With R. C. Mason. High velocity vapor jets at cathodes of vacuum arcs. Phys. Rev. 37:779-80.

1932 With L. R. Ludwig. Backfire in mercury arc rectifiers. Trans. AIEE 51:92-104.

1933 With L. R. Ludwig. A new method for initiating the cathode of an arc. Trans. AIEE 52:693-98.

1936 The ignitron: A new mercury arc power converting device. Trans. Am. Electrochem. Soc. 69:399-414.

The ignitron. Electr. J. 33:267-72.

1937 With R. C. Mason. The experimental validity of Paschen's law and of a similar relation for the reignition potential of an alternating current arc. J. Appl. Phys. 8:619-21.

1938 With A. H. Toepfer. Cathode spot fixation and mercury pool temperatures in an ignitron. J. Appl. Phys. 9:483-84.

1939 With W. M. Brubaker. Experiments on the condensation rate of mercury vapor. Phys. Rev. 55:1147(A).

1940 With W. E. Berkey. Spark gaps with short time lag. J. Appl. Phys. 11:765-68.

1941 With W. E. Pakala. Arcbacks in ignitrons in series. Trans. AIEE 60:292-94.

1942 Energy and energy flow in the electromagnetic field. J. Appl. Phys. 13:512-18.

1950 Electromagnetic ponderomotive forces within material bodies. Proc. Natl. Acad. Sci. U. S. A. 36:485-97.

1951 Lines of force in electric and magnetic fields. Am. J. Phys. 19:87-90.

1955 Failure of the ionic centrifuge. J. Appl. Phys. 26:1283.

I sotope separation by ionic expansion in a magnetic field. Proc. Natl. Acad. Sci. U. S. A. 41:4541-57.

1957 The magneto-ionic expander isotope separator. J. Franklin Inst. 263:129-39.

1958 Hydromagnetic equations for two isotopes in a completely ionized gas. Phys. Rev. 112:1441-44.

Part IX

Bibliography

1. Joseph Slepian, Electrical Essays, *Transactions of the American Institute of Electrical Engineers*, vols. 68 - 70, (January 1949 - April 1952)

2. Wolfgang K. H. Panofsky and Melba Phillips, Classical Electricity and Magnetism, Addison-Wesley Publishing Company, Inc., Cambridge, Massachusetts (1953) See section 9-4. Here the primes on $\mathbf{j}_{true}$, **P** and **D** are missing.

3. Joseph Slepian, Electrical Essay, *Transactions of the American Institute of Electrical Engineers*, p. 613, vol. 68, (July 1949)

4. M. Faraday, On Lines of Magnetic Force, *Phil. Trans. Roy. Soc. London* 142, 25 (1852)

5. H. A. Wilson and M. Wilson, Electric Effect of a Magnetic Insulator Rotating in a Magnetic Field. *Proc. Roy. Soc. Series A.,* **89**, p. 99 (1913).

6. W. K. Röntgen, *Ann. d. Phys. XXXV*, p. 264 (1888) and *Ann. d. Phys. XL* p. 93 (1890)

7. A. Eichenwald, *Ann. d. Phys. XI,* p. 421 (1903) and *Ann. d. Phys. XIII* p. 919(1904)

8. Richard P. Feynman, Robert B. Leighton and Matthew Sands, The Feynman Lectures on Physics, The Electromagnetic Field, pp. 17-2 - 17-3, Addison-Wesley Publishing Company, Inc., Reading, Massachusetts (1964)

9. William F. Hughes and Frederick J. Young, The Electromagnetodynamics of Fluids, pp. 28 -33, John Wiley & Sons, Inc. New York (1966)

10.Hermann Biondi and Joseph Samuel, The Lense–Thirring Effect and Mach's Principle, *usearXiv:gr-qc/9607009* (July 4, 1996)

11. Ezra Newman and Allen Janis, Note on the Kerr Spinning-Particle Metric, *Journal of Mathematical Physics* 6 (6), pp. 915–917. (1965)

12. Ezra Newman K. Chinnapared, A. Exton, A . Prakash and R. Torrence, Metric of a Rotating, Charged Mass *Journal of Mathematical Physics* 6 (6), pp. 918–919 (1965)

13. Frederick J. Young, A Course in Computational Electrostatic Field Theory (CEFT), 274 pages, Trafford Publishing, www.trafford.com (2012)

14. PDE Solutions Inc., FlexPDE 6 Manual, 517 pages, see page 230 (2009)

15 Peter P. Silvester and Ronald L. Ferrari, Finite Elements for Electrical Engineers, 3rd edition, Cambridge University Press, 1966

16. C. T. Tai, Generalized Vector and Dyadic Analysis, IEEE Press, New York (1992)

17. S.J. Barnett, On Electromagnetic Induction and Relative Motion, *Phys. Rev.*. 35, 323
(1912)

18. Albert Einstein, Zur Elektrodynamik bewegter Körper (On the Electrodynamics of Moving Bodies), *Ann. d. Phys.* XVII, p. 891, (sept., 1905)

19. Stanley W. Angrist, Direct Energy Conversion, Allyn & Bacon, 488 pages (1971)

20. J. A. Shercliff, The Theory of Electromagnetic Flow-Measurement, Cambridge University Press (1971)

21.Frederick J. Young editor of Magnetogidrodinamicie Metodi Polcyeniya Elektrikoi Energii, Trudi Institute Inshchenerov Po Elektrotechniki i Radio Elektroniki, Vol. 56, No.10, pp. 1-255, (1968)

22. Frederick J. Young, Magnetohydrodynamic Blanket Scaling in a Toroidal Fusion Reactor", Proceedings of the First Topical Meeting on Technology of Controlled Nuclear Fusion, ANS Conf-740402-PL, pp. 699-710.

23. Frederick J. Young, Magnetohydrodynamics Test of a One-Sixth Scale Model of a CTR Recirculation Lithium Blanket, TANSAO 21, pp/ 37-38 (June 1975)

24. Albert J. Martin and Frederick J. Young, Confinement magnétique des réactions thermonucléaires, Le Journal de Physique et le Radium, Vol. 20, pp. 1-4, (1959)

25. Thürlemann, B. Methode zur elektrischen Geschwindigkeitsmessung in Flüssigkeiten. (Method for electrical speed measurement in liquids.) Helv. Phys. Acta 14 (1941), pp. 383-419 3

Index

Printed in the United States
by Baker & Taylor Publisher Services